Resistance and Liquid-in-Glass Thermometry

Springer

Singapore
Berlin
Heidelberg
New York
Barcelona
Budapest
Hong Kong
London
Milan
Paris
Tokyo

Resistance and Liquid-in-Glass Thermometry

Editor

Robin E. Bentley

Robin E. Bentley
National Measurement Laboratory
CSIRO Telecommunications and Industrial Physics
P O Box 218
Lindfield
New South Wales 2070
Australia

Library of Congress Cataloging-in-Publication Data

Handbook of temperature measurement / edited by Robin E. Bentley
    p.   cm.
Includes bibliographical references
ISBN 9814021121 (set)

1. Temperature measurements--Handbooks, manuals, etc.
2.Hygrometry--Handbooks, manuals, etc. 3. Thermometers--Handbooks, manuals, etc. 4. Thermocouples -- Handbooks, manuals, etc. I. Bentley, Robin E. II. v. 1. Temperature and humidity measurement   III. v. 2. Resistance and liquid-in-glass thermometry   IV. v. 3. The theory and practice of thermoelectric thermometry.
QC271 .H276 1998
536/.5/0287 21
    98-28324
    CIP

ISBN 981-4021-10-5 (Volume 2)
ISBN 981-4021-12-1 (set)

© Springer-Verlag Singapore Pte. Ltd. 1998
Printed in Singapore

Typesetting: Camera-ready by editor
SPIN 10688208 (Volume 2)   5 4 3 2 1 0
SPIN 10688224 (set)   5 4 3 2 1 0

Preface

Temperature is one of the most widely measured physical quantities. Its accurate measurement is essential for the safe and efficient operation and control of a vast range of industrial processes. This book, one of three volumes in the Handbook of Temperature Measurement, is intended to give the reader an understanding not just of the appropriate techniques and instrumentation, but also of the underlying principles of measurement and sensor design. A discussion of the international framework for measurement traceability and the temperature scale is also given. Despite this, the emphasis is most focussed on the practical aspects of thermometry, such as calibration techniques and the identification and minimisation of error sources.

The books grew out of the biennial "Temperature Measurement Course" run by the Australian National Measurement Laboratory (NML). Started in the 1940's, the 3-5 day courses are given to participants from industry and laboratories involved in metrology and research, from Australia and the Asia-Pacific region. Each of the chapters is written by NML staff lecturing at this course, and expert in their field.

The Australian National Measurement Laboratory is a National Facility within the CSIRO (Commonwealth Scientific and Industrial Research Organisation), Division of Telecommunication and Industrial Physics. Established in 1938, it is located in Sydney and has responsibility for the maintenance and dissemination of the national units of measurement, for overseeing national measurement traceability and the international traceability of Australia's national standards. The temperature standards group within NML has played a leading role in international thermometry research, conducts industrial research projects and takes an active role in assisting industry in industrial measurement problem solving. This close contact with industry is reflected in the practical approach to thermometry given in these books.

Barry D. Inglis
Director
National Measurement Laboratory, Australia
May 1998

Contents

John J. Connolly

List of Tables

Chapter 1

Standard Platinum Resistance Thermometry

John Connolly

1.1 Introduction

Temperature

Our senses are able to detect the degree of hotness or coldness of objects around us and this introduces the concept of temperature. However, for its definition and that of any related temperature scale we need the study of thermodynamics and the concept of thermal equilibrium.

The zeroth law of thermodynamics states that if bodies A and B are both in thermal equilibrium with a body C, then they are in thermal equilibrium with each other. The parameter that describes the thermal state of a body, a parameter that must be equal to the same parameter in another body for them to be in thermal equilibrium, is called temperature. For a more detailed account/definition of 'temperature', of how it arises in thermodynamics and of the principles of temperature measurement, see reference [1].

Temperature is an intensive property of a material, and its measurement, of necessity, involves observing its variation with some other property more amenable to scaling. Obviously, the chosen property and thermometric substance should have a monotonic and predictable temperature dependence. Also, it is desirable that the dependence should be derivable from fundamental considerations. However, it is only in simple systems that theoretical relationships are known with any accuracy. An example is the constant-volume gas thermometer, where the temperature of the thermometric material is related to the pressure by consideration of the average kinetic energy of the gas molecules. This is the basis of a theoretical thermodynamic temperature scale for low and moderate temperatures. Another relationship that has firm theoretical backing, concerns the energy and wavelength distribution of thermally-generated radiation.

Thermodynamic temperature

Many of the fundamental physical constants include a temperature dimension, and the magnitude of the temperature unit, the kelvin (K), is set by assigning to the state of equilibrium between liquid water, water vapour and ice, the *water triple point* (WTP), the numerical value of 273.16 K. This means that the water triple point is the fundamental reference state for all temperature measurement. The rest of the thermodynamic scale is derived using methods described below, but the thermometers that measure thermodynamic temperatures are difficult to use, and measurements with uncertainties less than 20 mK are very difficult. Much work has been done at various national laboratories in recent times to define thermodynamic temperature and, in particular, to assign values to various reproducible physical transitions, such as normal freezing points of pure metals. Work on gas thermometry at the National Measurement Laboratory (NML), Australia [2] and at the US National Institute of Standards and Technology (NIST) [3, 4] over the few decades prior to 1990, has shown up considerable errors in previous thermodynamic temperature measurements from very low temperature up to 933 K. Overlapping this temperature range, there have been a number of total-radiation [5] and photoelectric-pyrometer [6, 7] experiments that have both confirmed the gas thermometry results and extended thermodynamic measurements to 1235 K.

However, even with these improved results in thermodynamic temperature measurement, the uncertainties obtainable are much greater than the sensitivities and reproducibilities that can be obtained from other empirical, rather than theoretical, thermometers. In most scientific temperature measurement, as well as virtually all industrial temperature measurement, the requirement is for either temperature repeatability or the measurement of temperature differences. For example, in the testing of the viscosity of oils, a temperature precision of 0.01 K is called for, but this is only so that different batches can be compared under identical conditions, i.e., it is the same temperature that is required, but its value in absolute terms is unimportant. This gives rise to the concept of an empirical or 'practical' temperature scale.

International temperature scales

A practical temperature scale is based on a series of reference temperatures defined by assigning numerical values to various reproducible physical transitions, mainly changes of state. The values are chosen to represent the best approximations to the thermodynamic values as known at the time, but, as improved experimental methods have led to revisions of these values, so also has the practical scale undergone several updates.

International Temperature Scale of 1990

The present scale is known as the *International Temperature Scale of 1990* (ITS–90) and replaced the previous International Practical Temperature Scale of 1968 (amended edition 1975), IPTS–68, in January 1990. ITS–90 is the fourth scale since the initial one was established by international agreement in 1927. The full text of ITS–90 is given in Chapter 9.

As stated above, the scale is set up by assigning values to some physical changes of state. For temperatures between these defined temperatures, a number of interpolating instruments are used, and the forms of the interpolating equations are specified. The constants in these equations are determined mostly from measurements at the defined or calibration points.

This chapter deals with interpolation instruments that use the relationship between electrical resistance of platinum and temperature as the basis for thermometers. Before discussing these relationships it is worthwhile looking briefly at the physical processes giving rise to the temperature dependence of resistance and then setting out some of the terminology used.

Elementary theory of temperature dependence of resistance

Although it is not possible to develop precise formulae for the temperature dependence of the resistance of Pt, the dependence exhibits an extremely stable and well characterised form and Pt wire is used as the sensor of very accurate, though empirical, thermometers. It is possible to gain some understanding of the phenomenon by consideration of the free electron model of electrical conduction.

The flow of electrons in a metal under the influence of an imposed electrical field is related to a bulk property of the metal, its *resistivity*. The resistivity, ρ, is a measure of the impedance to the motion of electrons in an electrical field and is governed by their interaction with the variations in the electric fields produced by vibrations of the metallic crystal lattice. These vibrations are temperature-dependent and the simple theory leads to:

$$\rho \propto T, \tag{1.1}$$

where T is the absolute temperature.

The resistance, R, of a wire is related to its resistivity, ρ, by

$$R = \rho \frac{l}{A}, \tag{1.2}$$

where l is the length of the wire and A, its cross-sectional area. As the wire temperature is increased the length increases by thermal expansion, but so does the cross-sectional area and the overall effect of the thermal expansion is the introduction of a second-order negative term,

$$R = a + bT + cT^2, \tag{1.3}$$

where, for metals, b is positive and c is negative. Unfortunately, theory cannot predict the magnitude of the coefficients. As the temperature is increased, further processes that impede electron flow begin to appear. Of these, the interaction of lattice defects or vacancies with the electrons is probably important and the concentration of such defects increases exponentially with temperature. Current theory, even with the introduction of quantum mechanics, cannot predict these effects with any degree of precision. An equation of the form of equation (1.3) was found to be suitable for the interpolation formula for platinum-resistance thermometers (PRT) when the first international temperature scale was set up, at least for temperatures above 0°C. However, as knowledge of the dependence of resistance on thermodynamic temperature has improved it has been found necessary to add further, higher-order terms to equation (1.3). For example, ITS–90 has a polynomial of ninth order to describe the resistance-versus-temperature relationship of Pt from 0 to 962°C.

1.1.1   Definitions

1. SPRT is a standard platinum-resistance thermometer satisfying the requirements of ITS–90.

2. T_{90} is the ITS–90 temperature in K.

3. t_{90} is the ITS–90 temperature in °C.

4. R_t is the resistance of a PRT at t°C.

5. R_0 is the resistance of the PRT at 0°C.

6. WTP is the water triple point defined as having a temperature of 273.16 K or 0.01°C.

7. R_{tp} is the resistance of the PRT at the water triple point.

8. W is the ratio of the resistance R_t to R_{tp}:

$$W = \frac{R_t}{R_{tp}}. \tag{1.4}$$

9. $W_r(T_{90})$ is the value of an ITS–90 reference function at T_{90} (see below).

1.2 ITS–90 reference functions

The standard platinum resistance thermometer (SPRT) is the ITS–90 inter-polation instrument from ∼13.8 K to ∼962°C. To accurately represent the resistance-temperature relationship of any given SPRT with a polynomial the required degree would exceed the number of suitable fixed points. Hence, the behaviour, and thus the calibration, of SPRT's on ITS–90 is expressed as two components. The first is a reference function that embodies the intrinsic functional dependence of R on T for all good-quality SPRT's, and the second is a deviation function to express the difference between the calibration of an individual PRT and the reference function. Fortunately, when given in terms of resistance ratios the deviations are relatively small and are expressible as low-degree polynomials.

There are two ITS–90 reference functions (Chapter 9), a twelfth-degree polynomial for the temperature range ∼13.8 K to 273.16 K, and a ninth-degree for the range 0°C to 961.78°C (the freezing point of Ag). The latter is:

$$W_r(T_{90}) = C_0 + \sum_{i=1}^{9} C_i \left[\frac{T_{90} - 754.15}{481} \right]^i. \tag{1.5}$$

An inverse function, equivalent to equation (1.5) to within 0.13 mK, is also defined:

$$T_{90} - 273.15 = D_0 + \sum_{i=1}^{9} D_i \left[\frac{W_r(T_{90}) - 2.64}{1.64} \right]^i, \tag{1.6}$$

where values of the coefficients C_0, C_i, D_0 and D_i are set out in Table 9.4.

The reference function for the temperature region from 13.8033 to 273.16 K is given in Chapter 9 (as equation (9.9), with its coefficients in Table 9.4). Apart from a brief mention of temperatures below 0°C, this chapter will deal only with measurements above 0°C. Details of thermometry in the cryogenic range are given elsewhere [8].

The deviation equations for various overlapping temperature ranges are given below.

1.2.1   Deviation equations

The deviation function covering the temperature range from $0\,°\mathrm{C}$ to the freezing point of Ag is

$$W(T_{90}) - W_r(T_{90}) = a\left[W(T_{90}) - 1\right] + b\left[W(T_{90}) - 1\right]^2 + c\left[W(T_{90}) - 1\right]^3$$
$$+ d\left[W(T_{90}) - W(660.323\,°\mathrm{C})\right]^2. \qquad (1.7)$$

Although equation (1.7) applies to the above range, it is split effectively into two sections. For temperatures below the freezing point of Al, $(660.323\,°\mathrm{C})$, $d = 0$ and the values of the other coefficients are determined from the measured deviations from $W_r(T_{90})$ at the freezing points of Sn, Zn and Al. For temperatures above the freezing point of Al the same coefficients a, b and c are used, as for lower temperatures, but the value of d is obtained from the measured deviation of $W(T_{Ag})$ from $W_r(T_{Ag})$. As will be seen later in this chapter, the design requirements for SPRT's to be used at high temperatures are different to those for low temperature use. Also the changes that commonly occur in the calibrations of SPRT's after temperature excursions above about $550\,°\mathrm{C}$ would severely limit the precision obtainable at lower temperatures. For these reasons ITS–90 has a number of sub-ranges that cover parts of the full range outlined above.

Convention for the coefficients of deviation equations

The ITS–90 does not differentiate between the coefficients in the deviation equations for the various ranges and sub-ranges, even though they are normally different. This leads to some confusion when an SPRT report gives the calibration coefficients for several ranges. As a result a convention has arisen that ascribes subscripts to the various ranges. Table 1.1 lists the subscripts to the deviation equation coefficients, a, b, and c, for the ranges and sub-ranges of ITS–90.

NML now follows this convention in reports of calibration of SPRT's and also when reporting the calibration coefficients of some digital resistance thermometers that employ the ITS–90 interpolation equations (see Chapter 4).

From $0\,°\mathrm{C}$ to the freezing point of Al ($660.323\,°\mathrm{C}$)

As mentioned above, the deviation equation is equation (1.7) with $d = 0$, and the calibration is carried out at the Sn, Zn and Al freezing points. The subscript 7 is used to differentiate the deviation equation coefficients a, b and c for this range.

Table 1.1: Deviation equation coefficients for the SPRT sub-ranges of ITS–90

Subrange	Calibration Points	Coefficients
83.8058 K to 273.16 K	Ar, Hg, WTP	a_4, b_4
-38.8344°C to 29.7646°C	Hg, WTP, Ga	a_5, b_5
0°C to 29.7646°C	WTP, Ga,	a_{11}
0°C to 156.5985°C	WTP, In	a_{10}
0°C to 231.928°C	WTP, In, Sn	a_9, b_9
0°C to 419.527°C	WTP, Sn, Zn	a_8, b_8
0°C to 660.323°C	WTP, Sn, Zn, Al	a_7, b_7, c_7
0°C to 961.78°C	WTP, Sn, Zn, Al, Ag	a_6, b_6, c_6, d

From 0°C to the freezing point of Zn (419.527°C)

The deviation equation for this range is reduced to a quadratic function:

$$W(T_{90}) - W_r(T_{90}) = a_8 \left[W(T_{90}) - 1\right] + b_8 \left[W(T_{90}) - 1\right]^2, \qquad (1.8)$$

where the coefficients a_8 and b_8 are obtained from resistance-ratio deviation measurements at the freezing points of Zn and Sn. This is probably the range of main interest for SPRT's, and the calibration points are the same as were used for an SPRT calibration on IPTS–68. Hence, calibrations for the 0 to 630.74°C range on IPTS–68 can be converted readily to ITS–90, at least for the range from 0 to 420°C.

From 0°C to the freezing point of Sn (231.928°C)

The deviation equation for this range is the same as for the previous sub-range, equation (1.8), but the calibration coefficients, a_9 and b_9, are obtained from measurements at the Sn and In freezing points.

From 0°C to the freezing point of In (156.5985°C)

The deviation equation for this range is reduced to a linear function:

$$W(T_{90}) - W_r(T_{90}) = a_{10} \left[W(T_{90}) - 1\right], \qquad (1.9)$$

and only a single-point calibration, at the In freezing point, is required (apart for the necessary water triple point measurement to determine the ratio $W(T_{In})$).

From 0°C to the melting point of Ga (29.7646°C)

The linear deviation function, equation (1.9), is also used for this sub-range, with the coefficient a_{11} determined from a ratio measurement for the Ga melting point. Since the solid volume for Ga is greater than that of its liquid, it is undesirable to use its freezing point as the reference temperature, as explained on page 20.

From the triple point of Hg (-38.8344°C) to the melting point of Ga (29.7646°C)

The deviation equation for this range is a quadratic:

$$W(T_{90}) - W_r(T_{90}) = a_5 \left[W(T_{90}) - 1\right] + b_5 \left[W(T_{90}) - 1\right]^2,$$

(1.10)

where the coefficients a_5 and b_5 are obtained from resistance-ratio measurements for the triple point of Hg and the melting point of Ga. This sub-range requires different reference functions from the other ranges to obtain the temperature, depending on whether the latter is above or below 273.16 K. Equation (1.5) is for use above 273.16 K and the reference function for temperatures below 273.16 K is equation (9.9) in the text of ITS–90 (Chapter 9).

1.2.2 Uniqueness

The term *uniqueness* refers to the ability of different SPRT's, each calibrated in accordance with the procedures contained in ITS–90, to give the same temperature values for the same temperature states. In practice, the scale is realised in the range 0 to 420°C with a precision and uniqueness of about 0.0005°C, but it is difficult to separate the effects of uncertainty of realisation of the calibration fixed points from inherent non-uniqueness in the way the scale is formulated. The possibility of non-uniqueness of ITS–90, arising because of the use of deviation equations as well as the overlapping of ranges, is still being investigated, but it is thought to be lower than the practical realisation uncertainty of a temperature determination in any part of the SPRT range.

The calibration of SPRT's is discussed in detail in section 1.4.

1.3 Standard platinum-resistance thermometers

In section 1.1 it was stated that temperature could be measured by relating the variation of a scalable property of a material to temperature. One such property is the electrical resistance. In the case of pure metals, resistance is roughly proportional to temperature and the relative sensitivity to a change in temperature is of the order 0.4 to 0.6% per °C. Many metals can be, and have been, used as resistance thermometers, although, Pt is clearly superior to virtually every other metal. The properties that recommend it are:

High resistivity	– compact sensor
Chemically inert	– corrosion resistant
Oxidation resistant	– lead attachment is simplified
Can be annealed	– easily drawn into small diameter wire

H. L. Callendar [9, 10] developed the first standard PRT in the late nineteenth century, but the basis of current practice comes from the 1940's designs of C. H. Meyers [11] of NIST. He used mica as the former upon which to wind the sensor coil of platinum. Modern standard instruments follow similar winding patterns, but the former is now usually pure quartz glass. Mica has very good electrical insulation properties, but occurs in thin sheets or leaves. These can contain free water between the leaves, as well as chemically combined water in the crystal structure. This limits the high-temperature usage, as free water causes electrical leakage that shunts the sensor.

Much work has been done in recent years to develop SPRT's capable of use to about 1000 °C and the success of this has enabled the extension of the range of the SPRT as an interpolation instrument for ITS–90 to the Ag freezing point.

The structure of SPRT's

ITS–90 requires that the highest-purity Pt be used for the sensor and this is ensured by specifying the following resistance-ratio requirements,

$$W(T_{Ga}) \geq 1.11807$$

$$W(T_{Hg}) \leq 0.844235$$

These requirements are equivalent to $W(100°C) \geq 1.3925$.

The platinum wire is formed into a compact coil wound on a former in such a way as to support the coil without introducing any strain. Fully annealed Pt is very soft and many designs have been tried in an attempt to achieve the necessary support. Some of these are illustrated in Figures 1.1 and 1.2.

The R_{tp} value sets the ohm-sensitivity of the thermometer (to changes in temperature). SPRT's often have an R_{tp} value of about 25.47 Ω, which, because of their relative sensitivity of 0.4%/°C, translates to about 0.1 Ω/°C

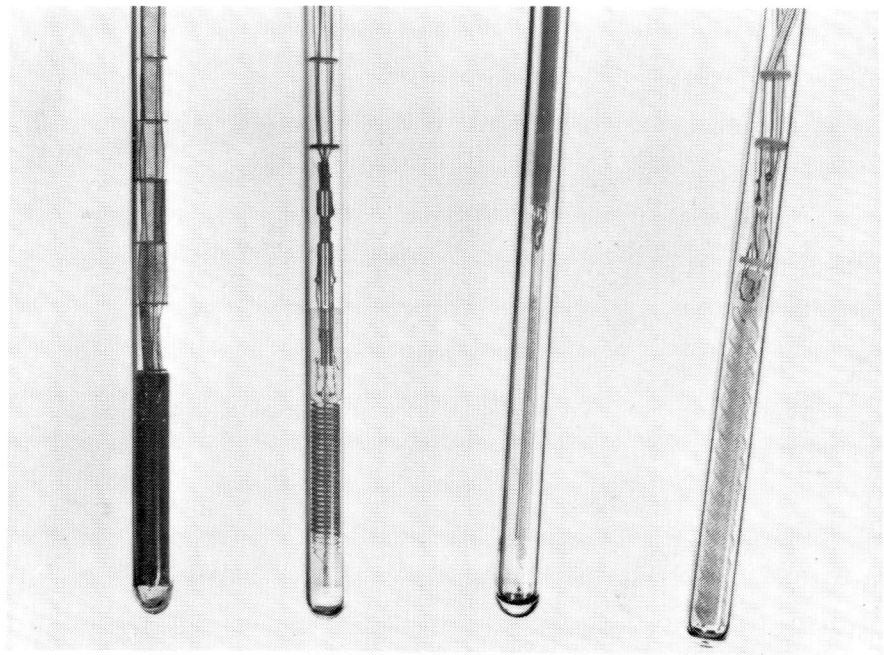

Figure 1.1: Examples of standard platinum-resistance thermometers.

at ambient temperatures, convenient for relating temperature changes to measured resistance changes.

As for all precision resistance measurements, a four-lead wire configuration must be used to remove the effects of lead resistance. The two pairs are normally referred to as current and potential leads. A typical SPRT has a Pyrex or quartz-glass sheath 450 to 500 mm long and 6 to 7 mm outside diameter. The sensor is 40 to 50 mm long, non-inductively wound, and has Au or Pt leads insulated with glass or silica capillary tubing. The sheath is normally hermetically sealed, with a partial atmosphere of dry air or oxygen-argon. The leads are brought through a seal and are either connected to a socket or to flying leads. The sealing and internal atmosphere are very important, as the presence of moisture gives rise to leakage errors. The structure of typical SPRT's is illustrated in Figure 1.1.

SPRT's designed for high-temperature use

ITS–90 has extended the use of the SPRT as an interpolating instrument to cover the temperature range previously covered by the type S thermocouple. PRT's for use in this higher-temperature range are somewhat different in structure than those designed for lower temperatures [12, 13]. Experience has shown that the best material for the sensor former, as well as for the sheathing and internal insulators, is pure fused quartz. Even this material becomes a less-than-perfect insulator as the temperature is increased, but the shunting effect is reduced by using a lower R_{tp} value for the sensor. This, of course, reduces the sensitivity of the PRT (in $\Omega/°C$), so the value chosen for R_{tp} is kept high enough, usually at $0.25\,\Omega$, for it not to impose too severe a constraint on the precision of resistance-measuring instruments.

At higher temperatures the mobilities of any impurities are higher and contamination of the sensor Pt is always a possibility. The lower R_{tp} used can result in larger diameter wire being used in the sensor. This reduces the risk of contamination because of the advantageous surface-area-to-volume ratio. Figure 1.2 illustrates some currently available designs for high-temperature SPRT's.

A further problem with high-temperature PRT's has been reported. There is some evidence that volatile components from Ni-based alloys can diffuse in some way through the silica sheath of an SPRT at temperatures as low as 700°C [14]. It therefore seems necessary to protect high-temperature SPRT's from exposure to such materials by interspersing other substances that may reduce the diffusion. The authors of this investigation have claimed that a thin-walled Pt sheath over the silica thermometer-sheath removes the problem.

1.4 Calibration of an SPRT

Initial conditioning

Before any calibration procedure is carried out, the thermometer's sensor must be in a fully annealed state. During manufacture the Pt wire is partly annealed, so as to make it workable without being so soft that it is difficult to handle. The winding process will cause some work-hardening of the wire, but the final stage of manufacture normally consists of annealing for many hours at a temperature in excess of 450°C. This removes most strain and the wire is classed as being fully annealed. Higher annealing temperatures speed up the process, but also increase the risk of contamination through the migration of impurities. However, SPRT's that have been exposed to temperatures above 450°C must be put through a cooling procedure to remove crystal lattice defects. These procedures are outlined in section 1.6.

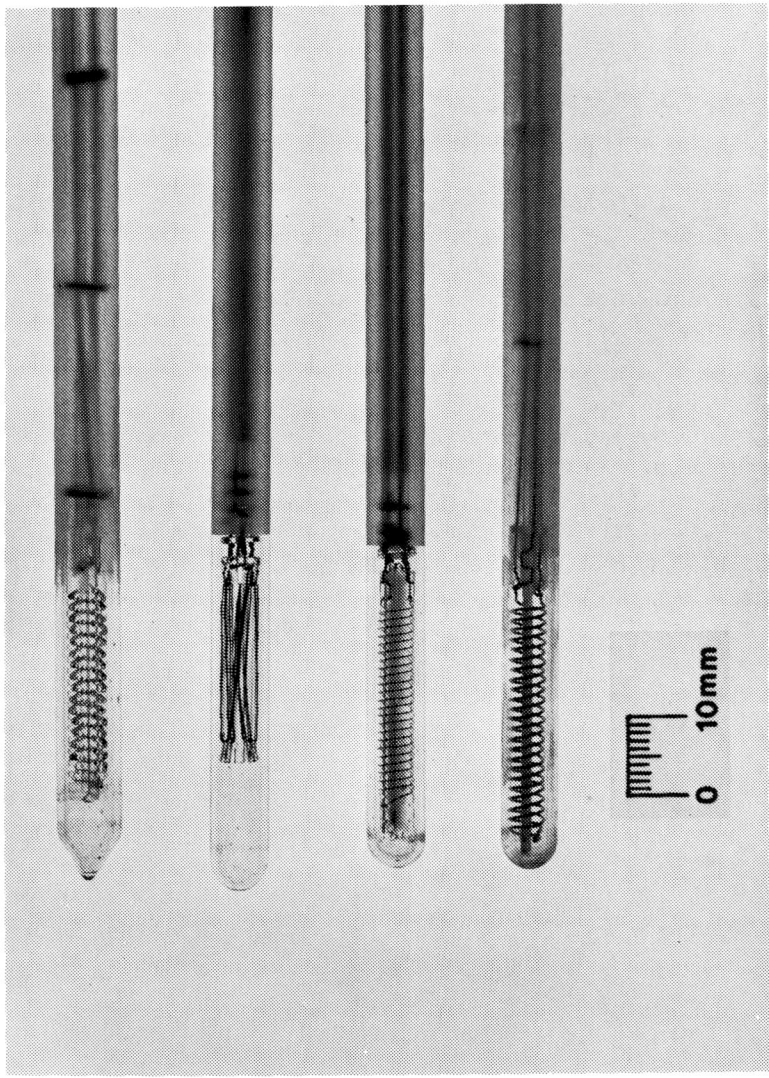

Figure 1.2: Examples of high-temperature standard platinum-resistance thermometers.

Measurement of R_{tp} for a PRT serves two major purposes:

1. It provides an easily obtained reference point for the determination of the resistance ratio, W, of the PRT. The PRT interpolation equations are always in the form of W as a function of temperature and, hence, are dimensionally independent of resistance. Thus, the calibration of PRT resistance-measuring instruments involves checking the linearity rather than a calibration in terms of the absolute ohm.

2. Stability in R_{tp} for a PRT is used as a measure of the continuing validity of its calibration. This arises from the observation that most small changes in R_{tp} cause proportional changes to resistances at other temperatures and, hence, W is unchanged.

Calibration

The other defined calibration points for SPRT's are the freezing points of pure In, Sn, Zn, Al and Ag as well as the melting point of Ga and the triple point of Hg. The techniques used to establish each of these points, together with those for some other useful secondary fixed points, are discussed below.

The resistance of the PRT is determined at the primary metal freezing points appropriate to the temperature range or sub-range, and each of these determinations is bracketed with water triple point measurements, so that the resistance ratios and differences from the reference resistance ratios can be calculated. The coefficients of the applicable deviation equation can then be obtained from the solution of simultaneous equations.

In a primary calibration, each fixed point will be repeated a number of times, but the process is time-consuming and the normal practice at NML is to carry out two sets of measurements at each point. An estimate of calibration uncertainty is obtained by two separate Cd freezing-point determinations (and such other redundant fixed points as may be available for the temperature range), as well as by consideration of the R_{tp} stability over the complete calibration.

1.5   Fixed points

1.5.1   Primary fixed points

Water triple point

The solid, liquid and vapour phases of a substance can co-exist in isolation only at a unique temperature and vapour pressure. This thermodynamic state is known as the *triple point* and the temperature is called the triple point temperature or just the triple point. The water triple point is normally realised in a sealed glass cell, such as illustrated in Figure 1.3.

The cell is filled with highest purity water from which dissolved gas has been removed. The ice-water interface is established by freezing the ice uniformly around the central re-entrant thermometer well, using a refrigerant (dry ice or liquid nitrogen) in the well. At NML a dry-ice-refrigerated heat-pipe [15] is used to freeze the ice mantle, as this method allows the ice crystals to attain macroscopic sizes within a few hours. (Microscopic crystals of ice have high surface curvature, and this lowers the equilibrium temperature from that of a planar surface by as much as some mK.) The water vapour phase is automatically established in the space above the water surface at a fixed pressure of 610.5 Pa.

In use, a so-called *inner-melt* water–ice interface is formed by inserting a rod into the thermometer well, thus melting back some ice from it, and placing some clean distilled water in the well to aid thermal transfer. The cell itself is used and maintained in an ice bath. The bath is, of course, 0.01 °C lower in temperature than the triple point cell, so that the cell is in a freezing environment. At NML the cells are placed in a polycarbonate cylinder in the ice bath, to reduce the freezing rate, so that the cell can be kept operational for many months with maintenance requirements limited to keeping the ice bath full and to draining off excess water.

Such cells realise the water triple point with an uncertainty of less than 0.1 mK, as evidenced by cell-to-cell and day-to-day variations. The occurrence of the most obvious faults, contamination and air leakage, can both be detected in the sealed cell. Ionic impurities increase the conductivity of the water, and there are a number of methods of measuring conductivity, involving external plates or coils and measurements of capacitance or inductance. The permanent gas pressure in the cell can be measured by the incorporation of a simple McLeod gauge. Other systematic errors that can occur in water triple-point cells are dealt with in section 1.7.

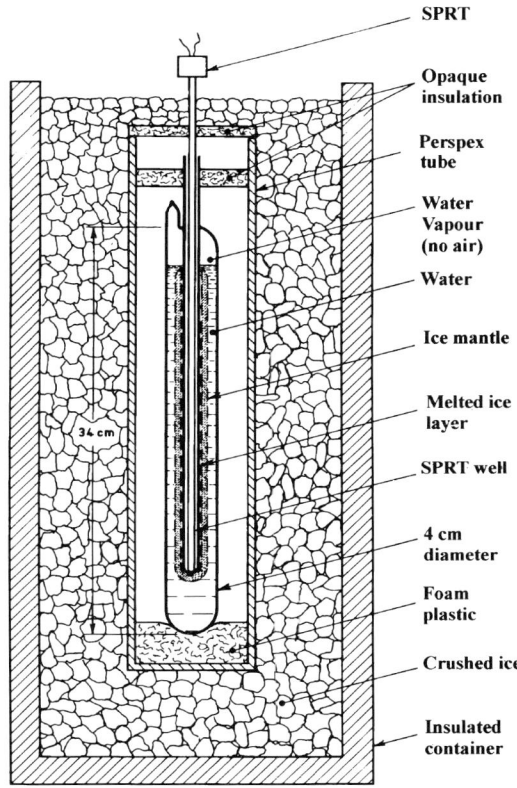

SPRT

Opaque
insulation

Perspex
tube

Water
Vapour
(no air)

Water

Ice mantle

Melted ice
layer

SPRT well

4 cm
diameter

Foam
plastic

Crushed ice

Insulated
container

34 cm

Figure 1.3: Water triple-point assembly.

Metal freezing points

The crucible assemblies, furnaces, and general techniques used for all the metal freezing points are very similar and follow from the investigations carried out by McLaren [16] many years ago. The metal is contained in a crucible, in most cases made of carbon, having an internal diameter of about 40 mm and a length of 230 mm.

Figure 1.4 illustrates a typical cell. The thermometer well can be of Pyrex, quartz or graphite. In most cases an inert atmosphere (argon) is used to prevent oxidation, but the pressure must be maintained at 1 atmosphere to conform to the normal freezing point conditions (Hg is normally realised as a triple point and hence is set up in a sealed cell at its own vapour pressure).

The major requirement for the furnace, also illustrated in Figure 1.4, is to provide a uniform-temperature zone around the carbon crucible. This

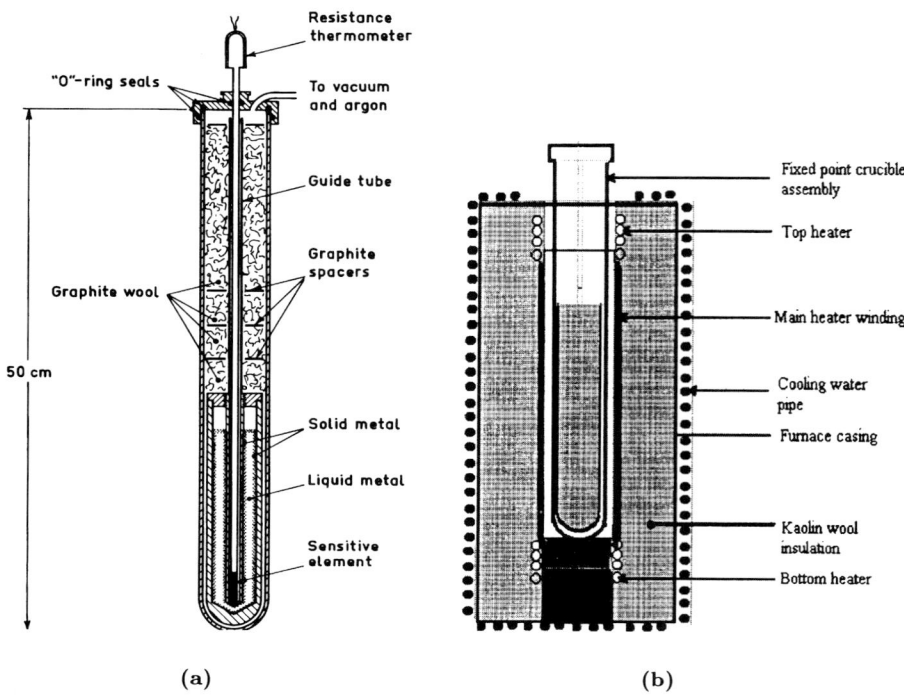

Figure 1.4: Schematic diagrams of a freezing point assembly, (a), and a freezing point furnace, (b).

is obtained by the use of either separately controlled heater zones or an appropriate heat-pipe liner(see section 8.3.3).

In order to obtain a stable freezing temperature the solid/liquid interface must be coaxial with the thermometer and grow radially inwards. The techniques used to nucleate the freeze and to control the growth of the interface differ from metal to metal, and are described below in more detail for each material. The most reliable techniques require a freezing time of some 1.5 to 2 hours, corresponding to an interface growth rate of about 0.1 mm/minute. Most impurities in the highly-purified materials used in reference points tend to be more soluble in the liquid phase than in the solid, so that the liquid phase tends to become enriched in impurities as the freeze progresses and the temperature becomes depressed towards the end of the freeze.

If the sample is melted slowly (say over 90 min.) immediately after the freeze, so as to minimise impurity diffusion effects, the melting range and melting temperature depression, when compared with the freezing temperature, give some quantitative estimate of the impurity levels present. This makes the procedure self-checking to some extent. A total impurity level of 1 ppm results in a melting range of less than 1 mK and 10 ppm gives a range of 10 mK, and a similar depression of the melting point. However, in the initial part of the freeze, the temperature is depressed by only a few mK below the freezing point of the pure material. The temperatures assigned to the defining fixed points dealt with in this chapter are set out in the table on page 25.

We now turn to specific fixed points.

Freezing point of tin

The techniques used to freeze tin differ somewhat from the others realised at NML because of the tendency of the liquid to supercool by 10 to 20°C or more before nucleation of the solid. If the sample, which contains about 1.2 kg of metal in the NML crucibles, is allowed to nucleate in the furnace, its latent heat is used up in raising the furnace temperature from that to which it had supercooled to the freezing point. To avoid this, a technique called *outside nucleation* is used. The sample is held 3 or 4°C above the freezing temperature to ensure that it is fully melted. Then, the furnace power (or set point) is reduced to lower the furnace temperature to about 5°C below the freezing temperature. The whole crucible assembly is lifted out of the furnace, when the monitor thermometer indicates the temperature is 1.5°C above the freezing point, and the temperature is tracked until the sample nucleates. The crucible is immediately returned to the furnace, which is then at about the freezing temperature. Hence, the latent heat has to raise the temperature of only the sample and crucible and not that of the furnace. The recovery *recalescence* is very fast, the thermometer being within 0.01°C of the freezing temperature in a minute or so.

Decanting samples part way through a freeze indicates that the initial nucleation is on the outside wall in a series of blobs, but these quickly amalgamate and the solid interface runs over the top surface and down the re-entrant thermometer well. Thus, as the sample reaches equilibrium, there are two solid/liquid interfaces, one near the thermometer well and the other on the crucible's inner wall, with liquid in between them. The thermometer records the temperature of the inner interface on the well. This interface will grow very slowly, as the heat flux is controlled by the other (outer) interface. The result is a very stable temperature and the thermometer is in a near-equilibrium environment.

A check melt is accomplished by first allowing the sample to cool to a few degrees below the freeze, to ensure that no liquid remains. The furnace power (or set point) is then raised to a value that will melt the sample in about 90 minutes. Sn points at NML contain commercially available tin with about 1 ppm impurity, and the freezing temperatures agree to 0.1 or 0.2 mK, with melting ranges of less than 1 mK for 90% of the melting time. The furnace used has separate top and bottom heaters that are adjusted to maintain a temperature uniformity of 0.01 °C/cm over a length of 500 mm, centred on the crucible.

Freezing point of zinc

The crucibles, assemblies, furnaces, and furnace-uniformity requirements used to realise the Zn freezing point are the same as those for the Sn point. The main difference lies in the techniques used to nucleate the sample and establish a symmetrical interface. Zn samples supercool only 0.01 to 0.02 °C, so the sample is allowed to self-nucleate in the furnace.

The procedure is, as far as the furnace is concerned, the same as for Sn, but immediately after the PRT has indicated nucleation it is withdrawn and two glass rods are inserted, in turn, into the thermometer well, each for about 45 seconds. The PRT is then replaced. This procedure, referred to as an *induced freeze*, establishes a second solid/liquid interface around the thermometer well.

Figure 1.5 shows typical freezing and melting curves for NML reference fixed points of zinc, tin and indium. The flat region of the freezing curve defines the fixed-point temperature and the quality of the material is assumed to be satisfactory if the melting range is less that 1 mK for at least 90% of the transition time.

Freezing point of aluminium

Al has a freezing temperature of 660.323 °C. It is only in comparatively recent times that 99.9999% pure aluminium has become commercially available. However, even after taking into account its very high reactivity and the need to maintain an argon protective atmosphere, with very low levels of water vapour, in the crucible assembly, the freezing point is very stable and reproducible [17] and, as a result, is now a defining point of ITS–90.

The techniques required to establish an Al freeze are different from those described above, as Al supercools about 1 or 2 °C. When the monitoring thermometer indicates a temperature about 0.25 °C above that of the freeze, the crucible assembly is lifted out of the furnace for about 30 seconds and replaced. The thermometer is then removed, a quartz-glass rod is inserted for about 45 seconds to chill a mantle on the well and the PRT is replaced. The freezing points of the Al samples at NML are reproducible to about 3 mK and the melting ranges are comparable.

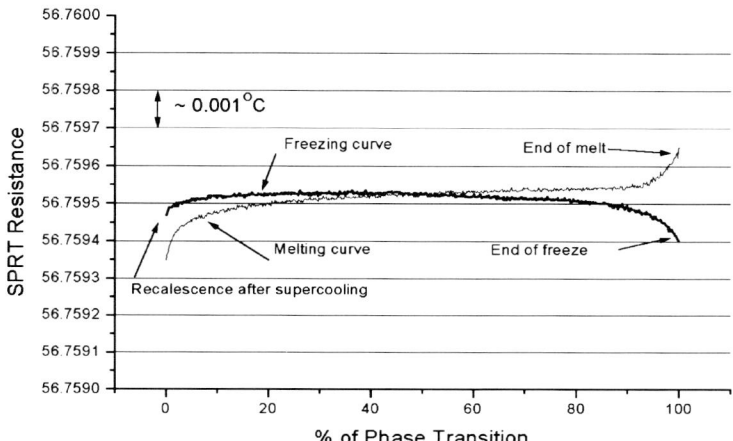

Figure 1.5: The graphs show the behaviour during a typical freezing point followed by a melting point. The metal is commercial, high-purity (99.9999%) cadmium, whose freezing point is 321.0685°C.

Freezing point of silver

SPRT's intended for use as ITS–90 interpolation instruments at high-temperatures require calibration at the Ag freezing point, a temperature within that part of the scale formally covered by the type S thermocouple. Considerable work has been undertaken at NML in studying the freezing and melting behaviour of Ag at a precision level suitable for use with SPRT's.

As a result, the Ag freezing point is available for the calibration of high-temperature PRT's, as required. Ag supercools about 0.1 °C, so the nucleation technique is similar to Zn. The thermometer is removed when the sample is about 0.2°C above the freeze and replaced after 15 seconds, since it cools very rapidly at these temperatures.

The Ag point, using 99.9999%-purity commercially-available material, has an ITS–90 assigned value of 961.78°C and is reproducible to better than 0.01 °C.

1.5.2   Sub-range primary fixed points

There are four temperature sub-ranges in ITS–90 covering the interval from about -39 to $232°C$ (see table on page 7), and to facilitate calibrations in these ranges three additional fixed points have been utilised. They are the freezing point of In, the melting point of Ga and the triple point of Hg. The points are useful for PRT's requiring a precise calibration over a limited temperature span, for example, around normal ambient temperatures, and for thermometers whose construction does not permit exposure to temperatures above that of the Sn freezing point.

Freezing point of indium

The In freezing point is realised in a similar way to Zn, using the *induced-freeze* technique. Since the latent heat of fusion of In is lower than that of the metals discussed so far, the recalescence after nucleation is not as dramatic in In. The supercool observed is normally about $0.07°C$. Some silicone oil is used in the thermometer well to aid heat transfer. Teflon containers can also be used and experience elsewhere [18] indicates that the supercool is then smaller. The lower density of the crucible aids recalescence.

Melting point of gallium

Ga has a number of properties that distinguish it from the other fixed-point metals. The first is that the density of the solid is less than that of its liquid phase. If, during freezing, the free surface of the liquid freezes over, subsequent solidification will impose a high pressure on the solid/liquid interface, which will alter its temperature. If the pressure is transmitted to the containing vessel walls they may fracture. This is similar to the problems that can occur with the water triple point, particularly when the mantle is being frozen. The second is that liquid Ga has a tendency to supercool by very large amounts: $80°C$ is not uncommon. As the melting point is $29.7646°C$ it can be very difficult to induce the liquid to nucleate and begin freezing [19].

These problems have been addressed in commercial equipment for establishing the Ga point, as a melting point reference temperature. The metal is contained in a Teflon and Nylon composite container. This can withstand substantial stresses as well as providing nucleation centres that reduce the supercooling to less than $10°C$. The temperature controller provided with the apparatus ensures that the sample freezes from the bottom upwards and melts from the top downwards. Thus, the cylindrical symmetry, centred on the PRT, occuring in other fixed points is not achieved in this case. However, the heat input to the sample during melting is controlled to produce a very long melting time, about 10 hours, and the melt is conducted under essentially adiabatic conditions. Heat flow down the PRT stem that would otherwise disturb the equilibrium conditions is avoided by adequate immersion and, in any case, the temperature difference between the fixed point and normal ambient temperature is less than 10°C.

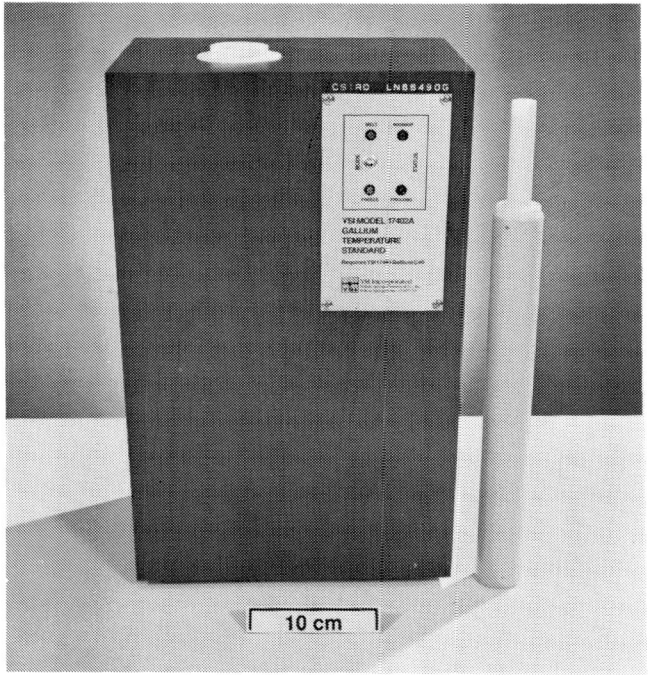

Figure 1.6: The commercially-available apparatus currently used at NML to realise the Ga melting point, showing the cell containing the Ga sample on the right.

The apparatus used at NML is illustrated in Figure 1.6; it reproduces the Ga melting point temperature to about ±0.3 mK. However, the oven is programmed to melt the gallium under close to adiabatic conditions over a time span of about 12 hours. This results in a single solid/liquid interface rather than the two co-axial interfaces that are set up in the freezing points described above.

The melting temperature range can be reduced by melting the gallium in a temperature-controlled stirred-water bath and establishing a second solid/liquid interface after the sample begins to melt, by placing an electric heater, set at about 45°C, in the thermometer well for about five minutes. The thermometer is also held at a temperature above the gallium point and replaced quickly so that a second interface is maintained. The melting range of the NML gallium point is then reduced to about 0.1mK.

Triple point of mercury

The Hg point is usually set up as a triple point, i.e., in a sealed container with only its own vapour occupying the space above the liquid surface. The NML apparatus contains the Hg in a stainless-steel vessel with a stainless-steel thermometer well. It is immersed in a stirred and refrigerated temperature-controlled alcohol bath and can be used in a freezing or melting situation. A copper rod cooled to liquid-nitrogen temperatures is used to induce a solid/liquid interface around the thermometer well in the same manner as used for zinc. The temperature appears to be reproducible to about ± 0.2 mK and the agreement between melts and freezes is of about the same order. An SPRT calibrated at the Hg triple point, the Ga melting point and the water triple point gives an extremely accurate realisation of ITS–90 in the ambient temperature range.

1.5.3 Secondary reference points

There are a number of convenient secondary reference points that are useful in the maintenance of ITS–90. A few of the more common ones are described.

Cadmium freezing point

This point is as reproducible as any of the above primary points and is used at NML as a check point for routine primary calibrations. The temperature has been determined as the mean value of a large number of measurements, the current value being $321.0685\pm0.0003°C$. The procedures and apparatus used are the same as for Zn. Cd has a high vapour pressure at its melting point and reacts with both oxygen and nitrogen. It is necessary to provide an atmosphere of argon, and as the vapour, nitride and oxide dusts are all poisonous, due care must be taken.

Antimony freezing point

Antimony was used as a primary fixed point for standard thermocouples on IPTS–68, although it only marked the upper limit of the SPRT temperature range and was not defined into the PRT calibration procedure. Its thermal properties result in it producing a poor thermometric fixed point and it has not been used as a secondary PRT point at NML for many years. Sb is toxic and can be very dangerous if spilled. It is still used for the calibration of secondary standard thermocouples.

Other phase-transition points

There are, of course, many other phase transitions that provide stable temperatures suitable for thermometric use. A listing of those whose temperatures are known is contained in the Supplementary Information for ITS–90 [20] with references to suitable methods of realisation.

1.6 SPRT conditioning

During manufacture, transportation and general handling an SPRT will be subjected to shock, which can cause structural changes to the sensor wire and affect its resistance. Some of the changes can be reversed by annealing at temperatures high enough to allow grain growth in the Pt crystal lattice and to promote the removal of defects. Too high a temperature may, however, cause contamination of the Pt by migration of impurities from the wire surface or surrounding structure.

The annealing process will cause a reduction in the value of R_{tp} for the PRT and, if the change is large, some alteration to the calibration coefficients will occur. It is normal practice to anneal most SPRT's at about 450 °C for a sufficient time to stabilise the R_{tp} value before proceeding with the calibration. An SPRT whose R_{tp} continues to rise during such anneals is regarded as an unsatisfactory instrument.

High temperature SPRT conditioning

The use of a PRT at temperatures above about 500 °C introduces a number of problems, and a different approach is required to maintain a reproducible R_{tp} value for the thermometer. Pre-conditioning before calibration will involve heating to temperatures above the maximum temperature of calibration, similar to the procedures outlined above. However, from about 450 °C upwards the concentration of lattice vacancies in the Pt lattice rapidly increases. If the sensor is cooled too rapidly, the excess lattice vacancies will be *quenched in*, resulting in an increase in R_{tp}. For this reason, any SPRT being removed from an environment whose temperature is above 450 °C is immediately placed in a furnace operating at about the same temperature and then the furnace temperature is ramped down at about 1 °C/min until below 400 °C. This procedure is carried out during initial annealing, during the calibration and whenever the PRT is subsequently used above 450 °C.

1.7   Errors in the use of SPRT's

There are two basic types of error that influence measurement—random and systematic errors, respectively referred to as 'type A' and 'type B' errors in the ISO guide to the expression of uncertainty in measurement [21], a topic also discussed in reference [22]. At this stage no mention will be made of any errors arising from the resistance-measuring equipment.

Sources of error in the use of the water triple point

Although the water triple point is a very reproducible fixed point, it is subject to errors that will cause a systematic offset to the temperature realised. Firstly, natural water contains a proportion of the heavier isotope of hydrogen, deuterium. There are variations in the concentration, depending on the actual source of the water and, also, the isotopic content is altered by distillation in the purification process during manufacture of the cell. D_2O has a triple point of about 4°C and, as the distillation process causes isotopic separation, the number of distillations has to be limited. NML cells all undergo two effective distillations.

The thermometer must be immersed sufficiently into the re-entrant well to prevent conduction errors (see page 25), but this means that the ice-water interface near it has the hydrostatic pressure of the water acting on it. This lowers the interface temperature by about $7.3\,\mu$°C/cm (allowed for in the most precise work).

The ice mantle and the water in a water triple-point cell are transparent to visible radiation and a PRT will 'see' bright illumination. The cell should be shielded from lights, particularly those using 'fluorescent' tubes.

Sources of error in the use of the fixed points

In the practical realisation of phase-transition fixed points, all the stipulated conditions for thermal equilibrium cannot be achieved. For example, as mentioned above, the required immersion depth of a PRT results in its equilibrium temperature being affected by the hydrostatic pressure of the liquid operating on the solid/liquid interface.

If significant, the hydrostatic-pressure effect should be corrected for using the Clausius/Clapeyron relation (see Table 1.2). There is a similar correction to be made if the external gas pressure acting on the system differs from the standard conditions. All the fixed-point temperatures are defined at zero hydrostatic pressure and, apart from water and other triple points, at standard atmospheric pressure.

Table 1.2: The temperatures of some defining fixed points and the hydrostatic and atmospheric effects on these temperatures.

Fixed Point	Temperature (°C)	Hydrostatic (mK/m)	Atmospheric (mK/atm)
Ice Point†	0	−0.73	−7.5
Water Triple Point	0.01	−0.73	−7.5
Hg TP	−38.8344	7.1	5.4
Ga MP	29.7646	−1.2	−2.0
In	156.5985	3.3	4.9
Sn	231.928	2.2	3.3
Cd†	321.0685	4.8	6.2
Zn	419.527	2.7	4.3
Al	660.323	1.6	7.0
Ag	961.78	5.4	6.0

† Not a defining fixed point on ITS–90

As can be seen from Table 1.2, the effects are small, but significant for the most precise measurements.

Three other sources of systematic error and the ways of identifying and counteracting them are discussed below.

Fixed-point impurities

As mentioned on page 15, the impurity content of a sample can be deduced in some cases from the melting and freezing ranges and the differences between them. Estimates can be made of the likely freezing-point depression on the basis of simple physical chemistry theory and allowances made, but the theory applies only to *ideal systems* and is not applicable to every case. It is better to reduce/avoid the problem by using a sufficiently pure sample, which, in most cases, is commercially available.

Errors arising from thermal conduction

The relative thermal resistances between the freezing interface and the PRT sensor, compared to that from the sensor along the PRT sheath and/or internal structure, will govern whether or not the PRT is close to thermal equilibrium with the solid/liquid interface. The heat flow along the thermometer can be minimised by using low thermal conductivity materials in its construction and by surface roughening or blackening the sheath and internal insulators to prevent thermal loss by light-piping. The thermal resistance across the thermometer well is reduced by ensuring that the well is a close, yet safe, fit on the PRT and, where possible, providing a thermal transfer fluid, for example, in a water triple point cell. However, care must be taken to ensure that any material in contact with the PRT does not act as a source of heat. Alcohol will absorb water vapour and there is a heat of mixing involved. A thermometer coated with colloidal graphite and exposed to oxygen will experience the heating effect of the carbon being oxidised, even at temperatures a little over 100°C.

The ultimate test to ascertain whether thermal conduction/radiation loss along the thermometer is a problem is to reduce the immersion of the PRT and see if the indicated temperature changes. This can be an extremely sensitive test in phase transitions, as the PRT should track the hydrostatic temperature gradient mentioned above. The freezing points maintained at NML do so fairly accurately, but with some slight variations that are attributed to movement of the effective sensing position of the PRT's sensor not being the same as the physical movement of the sensor.

Current-heating effect

An electric current is required in a PRT in order to measure its resistance. The current will dissipate power in the sensor and the resultant rise in temperature will be proportional to the power dissipated:

$$\Delta t \propto I^2 R. \qquad (1.11)$$

The dependence on the square of the current emphasises the need to keep the current as low as possible and also provides a method of correction of a measured resistance value to its zero-power value. The resistance is measured at two current values and it is a simple calculation to determine this correction. However, it should be remembered that it is the resistance ratio R_t/R_{tp} that occurs in the PRT interpolation formula and, as the heating effect is proportional to resistance, the ratio will tend to be independent of the current, provided the current causes only a small effect and the nature of heat transfer in the two resistance measurements is not too different.

1.8 Effect of calibration uncertainties

While the major sources of uncertainty in temperature measurements with SPRT's arise from the measurement environment and, perhaps, the resistance-measuring equipment, their calibration uncertainties also contribute. The uncertainties of realisation of the fixed-point resistance ratios result in uncertainties in the values of SPRT-calibration coefficients and estimates of the effects of these uncertainties at other temperatures can be calculated from elementary error analysis.

As an example of the procedures that can be followed, the calibration equation for the temperature range 0°C to the Zn freezing point will be considered. The calibration deviation equation is equation (1.8) and the two calibration coefficients a_8 and b_8 are determined from the differences between measured resistance ratios $W(T_{90})$ at the Sn and Zn fixed points and the corresponding values of the reference function $W_r(T_{90})$.

Alternatively, the calibration may be expressed as deviations given as a function of $W_r(T_{90})$, rather than of $W(T_{90})$, as it is in equation (1.8). It is then more amenable to error analysis, and after a little re-arranging:

$$W(T_{90}) - 1 = [W_r(T_{90}) - 1] + a\,[W_r(T_{90}) - 1] + b\,[W_r(T_{90}) - 1]^2,$$

where the coefficients a and b are not equal to a_8 and b_8 of equation (1.8). To simplify the algebra, make the following substitutions:

$$
\begin{aligned}
Y &= W(T_{90}) - 1 \\
\text{and} \quad X &= W_r(T_{90}) - 1, \\
\text{then,} \quad Y &= (1 + a)\,X + b\,X^2.
\end{aligned}
\tag{1.12}
$$

If the measured values of Y at the Sn and Zn fixed points are Y_1 and Y_2, respectively, and the corresponding reference values of X are X_1 and X_2, then it is a simple matter to obtain a and b from equation (1.12).

$$
\begin{aligned}
a &= \frac{Y_1\,X_2}{X_1\,(X_2 - X_1)} - \frac{Y_2\,X_1}{X_2\,(X_2 - X_1)} - 1 \\
\text{and} \quad b &= \frac{-Y_1}{X_1\,(X_2 - X_1)} + \frac{Y_2}{X_2\,(X_2 - X_1)}.
\end{aligned}
$$

The uncertainties in a and b arising from the uncertainty ΔY_1 in the measured resistance ratio at the Sn point and ΔY_2, at the Zn point, are given by

$$\Delta a = \frac{X_2}{X_1\,(X_2 - X_1)}\,\Delta Y_1 - \frac{X_1}{X_2\,(X_2 - X_1)}\,\Delta Y_2 \tag{1.13}$$

and

$$\Delta b = \frac{-1}{X_1\,(X_2 - X_1)}\,\Delta Y_1 + \frac{1}{X_2\,(X_2 - X_1)}\,\Delta Y_2. \tag{1.14}$$

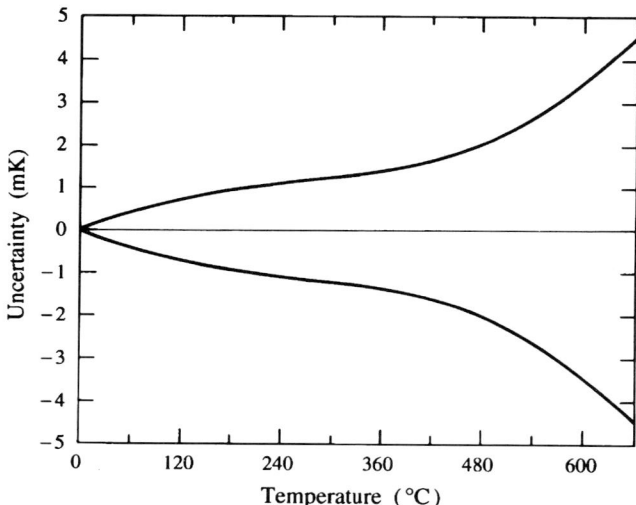

Figure 1.7: The likely uncertainty, ΔT, in an SPRT-temperature measurement at temperature T, arising from the equivalent of a 1 mK uncertainty in each of the Sn and Zn fixed points. The uncertainty is represented as confidence limits.

From these equations, it is relatively simple to calculate, using equations (1.6) and (1.7), the effect of Δa and Δb on subsequent temperature measurements. It is possible to draw error charts assuming various combinations of fixed-point errors. Figure 1.7 shows the uncertainty envelope for the equivalent of a 1 mK uncertainty in the measured resistance ratios at each of the Sn and Zn points, assuming that these components of uncertainty can be combined in a root-mean-square manner.

Similar analyses for the other temperature ranges can be made, but the algebra becomes more complex with the higher-order deviation equations.

Chapter 2

Industrial Resistance Thermometers

John Connolly

2.1   Introduction

Chapter 1 deals only with standard platinum-resistance thermometers (SPRT), but resistance thermometers in general are capable of quite reliable and economical industrial-quality temperature measurement and they are used very widely. The sensing material can be almost any pure metal, although, in practice, only copper, nickel and platinum are used to any extent. Australian standard specification AS2091 deals with resistance thermometers made of these materials. However, platinum proves to be the best material for most purposes, so the platinum-resistance thermometer is the most common form of this instrument and most national standard specification codes, as well as IEC–751 [23], are limited to platinum.

Not only must the sensors be designed to withstand shock and vibration, but the sheathing must be much more robust than for SPRT's. The stability of the thermometer and many of its characteristics will be influenced by the method of satisfying these requirements.

This chapter describes the types of structure that are necessary to make resistance thermometers suitable for use in an industrial environment. It also covers calibration procedures and discusses common sources of error in measurements of temperature made with an industrial platinum-resistance thermometers (IPRT).

Reference calibration point for IPRT's

The water triple point is the primary reference temperature in ITS-90 and is used as the reference point for the resistance ratios in the SPRT interpolation equations. However, the establishment, maintenance and use of water triple-point cells requires considerable practical expertise and the gain in measurement uncertainty is, in most cases, unnecessary for industrial thermometry. The ice point, 0.01°C lower than the water triple point temperature, provides a reference temperature with an uncertainty of about 0.002°C, if the procedures outlined in section 8.2.3 are followed.

If the ITS-90 interpolation equations are being used with an IPRT the resistance ratio at the ice point, W_{ip}, can be converted to W_{tp} with an uncertainty less than 0.002°C using the equation:

$$W_{tp} = W_{ip} \times 1.000039 \qquad (2.1)$$

2.1.1 Definitions

There are a number of terms and parameters that are commonly used in discussions of IPRT's and, although many have been introduced in Chapter 1, it is probably worthwhile setting them out again here.

- ITS–90 is the International Temperature Scale of 1990.

- IPTS–68 is the International Practical Temperature Scale of 1968, (amended edition of 1975) that was superseded by ITS–90 on 1 January 1990.

- SPRT is a standard platinum-resistance thermometer complying with the requirements of ITS–90.

- IPRT is an industrial quality PRT, not necessarily complying with ITS–90, but usable as a laboratory or transfer secondary-standard thermometer.

- RTD is an alternative for the term IPRT.

- R_t is the resistance of a thermometer at some temperature t°C.

- R_0 is the resistance at the ice point, 0°C.

- W is the ratio of R_t to R_0.

$$W = \frac{R_t}{R_0}$$

- W_{ip} and W_{tp} are respectively the resistance ratios at the ice point and the water triple point—see equation (2.1).

- α is the average slope of W-versus-t between 0 and 100°C:

$$\alpha = \frac{R_{100} - R_0}{100 R_0} \quad {}^\circ\mathrm{C}^{-1} \,,$$

where W_{100} is the resistance ratio at 100°C. Both α and W_{100} are used as purity criteria for the Pt of the sensor, and they are related by:

$$W_{100} = 1 + 100\alpha \,.$$

- $\alpha = 0.00385°\mathrm{C}^{-1}$ for sensors complying with IEC–751 [23].

- $\alpha \approx 0.003920°\mathrm{C}^{-1}$ for many secondary-standard PRT's.

- $\alpha \geq 0.003925°\mathrm{C}^{-1}$ for SPRT's.

2.1.2 Basic properties

Usually the temperature-sensitive resistor is made from a pure metal (platinum, nickel or copper), although some alloys, for example Rh Fe, are used for cryogenic temperatures and semiconductors are used in some applications (germanium, silicon and carbon for cryogenic temperatures and thermistors, diodes and some integrated-circuit devices near room temperature).

Desirable properties for the sensor material are:

- large change of resistance with temperature,
- linear change with temperature,
- stable resistance-temperature characteristics,
- high resistivity,
- corrosion resistance,
- resistance is stable to cycling of temperature,
- usable over a large temperature range,
- simple manufacture,
- robust construction (not affected by vibration) and
- low cost.

The properties of some materials are listed in Table 2.1. Of these, platinum is by far the most often used by manufacturers—its cost being an insignificant part of the total cost of a thermometer—because the properties of platinum are superior in most respects to those of other materials. Nickel is used for those applications where its large temperature coefficient is helpful, although modern detectors tend to have plenty of sensitivity to allow their use with platinum. The alloy of rhodium with 0.5% of iron provides a thermometer that has extra sensitivity down to $-272°C$ and can be used up to $400°C$. For this reason it is sometimes used in applications that span such a wide temperature range.

Table 2.1: Properties of materials used in the sensors of resistance thermometers.

Material	Temperature Coefficient (%/K)	Law	Resistivity	Range (°C)	Stability
Pt	0.4	Quadratic	Medium	-260 to $+1200$	High
Ni	0.6	Quadratic	Medium	-200 to $+350$	Medium
Cu	0.4	Linear	Low	-200 to $+200$	Medium
Rh 5% Fe	0.4	Quadratic	Medium	-272 to $+400$	High
Thermistor	-4	$A \cdot \exp(B/T)$	High	-200 to $+200$	Medium

Metals do not depart far from a linear relation between resistance and temperature, except at very low temperatures. Copper is almost perfectly linear, especially in the refrigeration and room-temperature regions, and this simplifies instrumentation and interpolation, although the non-linear response of platinum can be allowed for without excessive complication, in all but the cheapest of instruments. Even in the case of platinum, the departure from a linear curve between $0°C$ and $100°C$ is only about $0.4°C$ at $50°C$.

2.2 Standard specification codes

The standard specification code IEC–751 [23] has been incorporated into many national codes and describes the design, properties and testing procedures for IPRT's. However, the code is more concerned with interchangeability than with utilising the full measurement capability of IPRT's.

Interpolation equations

IEC–751 defines the temperature-resistance relationship for IPRT's in the temperature range 0 to $850°C$ by a simple quadratic equation:

$$R_t = R_0 \left[1 + At + Bt^2\right]. \tag{2.2}$$

For the range from -200 to $0°C$ a further term is used:

$$R_t = R_0 \left[1 + At + Bt^2 + C\left(t - 100\right)t^3\right]. \tag{2.3}$$

Values for the constants in these two equations are given in IEC–751, but the allowable variations must be determined from the appropriate tolerance relationship (see below). If these interpolation equations are compared with those given for SPRT's in Chapter 1 it will be realised that they give only approximations to t_{90}.

Tolerances

The temperature tolerances allowed in IEC–751 at a temperature, t (°C), for two classes of IPRT's are as follows:

- Class A: $\pm(0.15 + 0.002 \mid t \mid)$
- Class B: $\pm(0.3 + 0.005 \mid t \mid)$

It can be seen that for both classes the first term corresponds to the tolerance in R_0 and the second terms are 0.2% and 0.5%, respectively, of the temperature. These tolerances are somewhat tighter than those for commercial thermocouples.

The forms of the two interpolation equations (2.2) and (2.3) have been unchanged even with the changes that have occurred to the SPRT interpolation formulae as a result of revisions to the International Temperature Scale.

However, the tolerances are such that the changes introduced by revisions of the international temperature scales lie within the tolerance bands and the interpolation equations (2.2) and (2.3) are still valid at these levels of precision.

Figure 2.1 shows the allowable tolerances for the two classes of IPRT's covered by IEC–751. It can be seen that a single calibration at the ice point to determine R_0 reduces the uncertainty substantially, particularly at temperatures below 100°C.

Although there are situations where resistance thermometers with such broad tolerances are suitable, thermocouples are generally the more appropriate instruments for this class of work. IPRT's should be considered as an alternative or replacement for liquid-in-glass thermometers as it is only when a precision of better than 0.5°C is required that IPRT's become an economical choice. It is not possible to batch manufacture IPRT's with tolerances close enough to work to this level of precision (0.1% or better), so they have to be calibrated individually.

Another point worth noting is the possibility of confusion arising from current **terminology**: the words used to describe

(a) the sensing resistor whose characteristic changes with temperature and

(b) the assembly of the resistor plus leads and sheath.

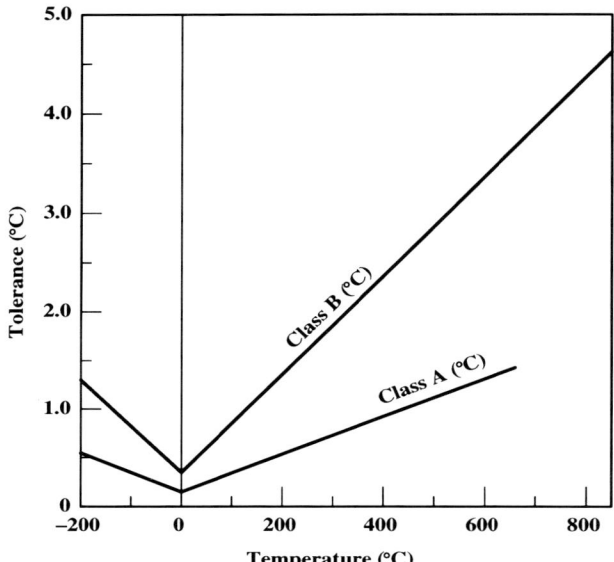

Figure 2.1: IEC–751 [23] tolerance values, ΔT, as a function of temperature, T.

Sometimes, the term *sensor* is used when referring to item (a) and in other instances to (b). IEC–751 uses *sensing resistor* for item (a) and refers to item (b) as a *platinum resistance thermometer*, a *resistance thermometer* or just a *thermometer*. However, in some cases, especially for direct-reading digital instruments, a *thermometer* includes the measuring device.

In this book, the AS2091 usage will be followed in that item (a) is referred to as the *element* or *sensor* and (b) as the *PRT* or *thermometer*.

Summary

Standard specification codes are mainly directed at the interchangeability criteria for PRT's used with temperature indicators or controllers. However, it should be noted that IEC–751 [23] also contains useful information on design parameters and testing procedures that can have application when considering more precise *transfer or laboratory standard PRT's*.

It should be realised that many IPRT's sold as meeting IEC–751 are capable, with individual calibration, of being used as laboratory standards.

2.3 Industrial platinum-resistance thermometers

IPRT's are used at various precision levels and in many different configurations in industrial measurement and control situations. Generally they are an alternative to thermocouples and offer some advantages in sensitivity, although they are more limited in temperature range. A typical industrial PRT will have an ice point resistance, R_0, of about $100\,\Omega$ and when used in a measuring circuit with a current of about $1\,\text{mA}$ has an effective sensitivity, in voltage terms, of about $400\,\mu\text{V}\,^\circ\text{C}^{-1}$. This is a factor of about forty greater than the sensitivity of a type R or S thermocouple. The main disadvantage of an IPRT is the need for lead compensation, which usually requires running three or four leads to the sensor rather than the two wires of a thermocouple. However, the true worth of PRT's is achieved after individual calibration, when their measurement capabilities become superior to most other thermometers.

IPRT's are viable replacements for laboratory mercury-in-glass thermometers in many applications from -30 to $+250^\circ$C, as a single PRT can cover this range with a precision approaching 0.01°C [24, 25], whereas a mercury-in-glass thermometer of similar sensitivity spans only 10°C, so that many such thermometers would be required. Also, being electrical instruments, PRT's can be used with recording or data-logging instruments for automatic reading, whereas liquid-in-glass thermometers are visual instruments.

Specifically designed IPRT's can be used to measure temperatures over 500°C with uncertainties an order of magnitude better than that obtained with thermocouples and can be used as secondary-standards to calibrate other devices or processes or to monitor critical thermocouples, to decide when maintenance or replacement is needed. For example, steam turbine generators are often provided with spare probe wells next to monitor and control thermocouples so that *in-situ* calibration checks can be made by PRT's.

While most users of IPRT's do not ever become involved in their construction, it is worthwhile considering the various elements that constitute a PRT so that sources of error can be identified and counteracted.

IPRT sensor construction

In the discussion on SPRT's the need for mounting the sensor's Pt wire in a strain-free manner was stressed (1.6). In the case of PRT's to be used in industrial or even most laboratory environments, such a structure would not survive the shock and vibration normally encountered. In many commercial PRT sensors the Pt is encapsulated in glass or a ceramic material.

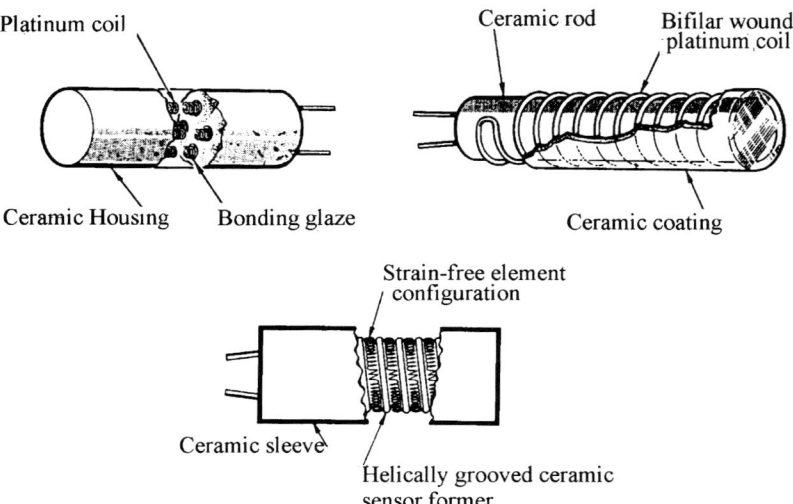

Figure 2.2: Examples of some designs of industrial PRT sensors.

Encapsulated Pt-wire sensors

In considering the design criteria of a platinum-resistance thermometer suitable for industrial applications the overriding consideration will be robustness, to enable it to cope with vibration and the generally unforgiving environment imposed in most applications. As stated above, the platinum wire in the sensor of an SPRT is wound in a configuration that is as strain-free as possible. This makes it rather susceptible to vibration and IPRT sensors are designed differently. The wire is supported in such a manner as to minimise vibration effects, and the great advance in general stability came with designs incorporating glass or ceramic materials whose thermal expansion coefficient matched that of platinum. This enables the wire to be well supported and restrained while not introducing *strain-gauge* effects. The degree of constraint of the wire varies according to the design used and Figure 2.2 illustrates some of the designs.

Most designs of this type have two leads sealed in, although a few have four lead junctions inside the ceramic. This simplifies the connection of the longer leads for potential-lead thermometers, reduces the risk of short circuits and makes the assembly less fragile. Leads that are sealed through the ceramic are often made from a Pt alloy rather than pure Pt, to better match the thermal expansion coefficient of the glaze used as a seal and to provide better strength. There is a trade-off. The thermal expansion coefficient of the ceramic cannot be matched to that of the platinum over the entire range of use of the PRT. Also the ceramic contains impurities, which will cause contamination of the Pt and degradation of the sensor's temperature coefficient, particularly when used at high temperature. Many versions of ceramic-encapsulated IPRT sensors suffer from a drop in R_0 after cooling below about $-40\,°C$. The sensor will retain this R_0 value provided it is not heated above about $100\,°C$, when it will revert to its former value. This hysteresis is probably due to a dimensional change in the ceramic.

These are the main reasons for the variability in stability and general performance of IPRT's. Those with the Pt encapsulated in glass also suffer from problems of electrical leakage above about $250\,°C$.

Even with the obvious disadvantages of having the Pt in such intimate contact with, normally, alumina-based ceramics, known to cause contamination, the encapsulated Pt-wire sensor is an attractive alternative to Hg-in-glass thermometers and thermocouples. There are sensors available at a moderate price that conform to the requirements of ITS–90 and perform almost as well as SPRT's, at least to $420\,°C$.

Thick-film sensors

The encapsulated wire sensors discussed in the previous section are relatively expensive as there is much labour involved in their manufacture.

There is another group of resistance sensors available where the resistor, usually Pt, is in the form of thick-film tracks laid down on a ceramic substrate. There are many forms of these *thick-film* sensors, some having total substrate areas of about $25\,mm^2$ and an active element area of $5\,mm^2$. They are mass produced by automated techniques and, although individually trimmed to the correct R_0 value, they are relatively cheap. They are robust, unaffected by vibration and, because of the film structure, they are completely non-inductive, making them reasonably insensitive to stray electrical fields. Like wire elements mounted in ceramics, they display instability when cooled below $-40\,°C$, which can be allowed for if required.

However, the intimate contact between the Pt and the substrate requires close matching of the coefficients of thermal expansion. It also exposes the Pt to contamination, by migration of impurities from the substrate during use at high temperatures.

For these reasons, thick-film sensors are normally designated to be used only to the Class B tolerances of IEC–751, and only perform at this level. They are widely used as temperature sensors in the domestic appliance and motor industries.

Rh Fe film sensors are made in forms similar to the Pt sensors. Resistances in the range 300 to 500 Ω are available and they have similar properties to the wire versions.

2.4   Thermistors

Thermistors are resistors made from mixtures of complex metallic oxides, and are semiconductors. The desired resistivity and temperature coefficients are controlled by heat treatment and composition. They have a high negative temperature coefficient of resistance, -3 to $-5\%\,°C^{-1}$ compared with a value of about $+0.4\%\,°C^{-1}$ for metals. The resistivity is normally much higher than for metals, enabling the fabrication of sensors with small dimensions and small response times. They can be obtained commercially with room-temperature resistances ranging from $100\,\Omega$ to $1\,M\Omega$ and can be useful for temperature measurement from about -200 to $300°C$. Their high sensitivity permits their use with inexpensive, direct-reading instruments that will maintain calibration for long periods if the thermistor is not mal-treated. As the lead resistance is normally small compared to the sensor's resistance lead compensation is unnecessary. Considerable design detail is available in thermistor data sheets for a variety of temperature measurement and control applications.

The resistance temperature curve is very non-linear and is usually represented by an equation of the form:

$$R = aT^c e^{\frac{b}{T}} \tag{2.4}$$

where T is the temperature in kelvin and c is a small positive or negative number close to zero. To a good approximation the relation reduces to:

$$R = ae^{\frac{b}{T}}. \tag{2.5}$$

Here b varies between 2000 and 6000 K, and a value of 3600 K is equivalent to a temperature coefficient of $-4\%\,K^{-1}$ at room temperature.

Manufacturers find it difficult to obtain reproducibility, even within the same batch of thermistors, because the exponential term changes the characteristic so rapidly. It is possible to check a batch at several temperatures and to select pairs having almost identical response curves over a limited temperature region. Such pairs can be used in simple and very precise temperature-difference measuring equipment, e.g., in heat-flow measurements. Nevertheless, because thermistors are not interchangeable without careful calibration, they are not used a great deal for industrial temperature measurement.

Because of their small volume, small surface area and large resistance, heating of the sensor by the measurement current produces very large errors in temperature unless very small currents are employed. Whereas currents of the order of a few mA are used for Pt sensors, a few μA are all that can be tolerated for typical high-resistance thermistors.

However, some suppliers specialising in temperature sensors produce surprisingly stable thermistors that can be used with a measurement precision of 0.01°C or better. Thermistors are used as sensors in some digital thermometers, particularly where small response times at normal ambient temperatures are required (see section 4.3).

2.5 Lead configurations

Four types of lead arrangement are used with PRT's, although in modern instrumentation only the potential four-lead style is of importance for secondary-standard and standard PRT's. However, all four will be discussed briefly and further detail will be given in Chapter 3.

Two-lead configuration

The simple two-lead connection to the sensor incorporates the resistance of the leads into the measurement. As a result there is no compensation and the lead resistance must be kept small relative to the sensor's resistance. Although this arrangement is used in industrial measurement, contributing uncertainties of about 1%, some form of lead compensation must be used if higher accuracies are required. The other approach is to use sensors with a high R_0, for example, thermistors, but careful consideration must be given to the magnitude of the measurement current to avoid current-heating errors. Two-lead PRT's are restricted to Class B tolerance in IEC–751.

Three-lead (Siemens) configuration

The Siemen's configuration uses a third lead to branch the measurement current at the sensor. This moves the junction of two of the arms of the bridge to one end of the PRT sensor and places one of the leads in the opposite arm to the other (see page 61). If the two leads actually in the bridge circuit are of equal resistance they will cancel each other, provided the bridge circuit is of an equal-ratio Wheatstone type. In practice, the uncertainty arising from a three-lead system is limited to about 0.01°C and it should be remembered that this system will only provide lead compensation if used in the correct bridge configuration.

Callendar four-lead compensating configuration

In the Callendar arrangement, the sensor has a two-lead connection and there are two additional leads of the same material, forming a loop beside the sensor's leads. The loop and the sensor are connected into opposite arms of a Wheatstone bridge and the lead resistance is compensated for in the same way as for a three-lead thermometer. Again, this configuration can only be used in an equal-ratio bridge network and has little application in modern instrumentation.

Potential (four-lead) configuration

The potential-lead configuration is used for all high-quality PRT's, either for use with precision resistance bridges or potentiometric resistance-measuring instruments. Then, the resistance of the leads has no significant effect (Chapter 3). Two leads, one for *current* and one for *potential*, are connected to each end of the sensor and the junctions uniquely define the sensor resistance.

2.6   Leads and lead insulation

The electrical circuit between the PRT sensor and its measuring instrument - *the thermometer's leads* - can be considered in two parts:

1. the leads within the *PRT*,

2. flying leads between the thermometer's head and the resistance-measuring instrument.

The metal of the **internal leads** can range from gold or platinum (or Pt Rh alloy) to silver or nickel-based alloys. At low temperatures Teflon or an electrical varnish is used to provide electrical insulation for the leads while still retaining their flexibility. IPRT's designed for use above 100°C usually have the leads insulated by a ceramic material. This may be in the form of multi-bore rod, individual tubes or woven ceramic fibre fitted around each lead wire. The material used should not contain substances that could cause contamination of the sensor Pt and should be able to be cleaned and dried before the thermometer is assembled. A common, convenient insulator is four-bore high-purity alumina rod.

The thermal conductance of leads and insulation should be minimised by the selection of appropriate materials, while ensuring that leads are not subject to corrosion and that their electrical resistance be as low as possible. Unfortunately, the thermal and electrical conductivities of metals usually go hand in hand, so a material with good electrical conductivity, chosen to reduce lead resistance, will tend to cause thermal conduction problems. Obviously the leads, insulation and method of lead attachment must all be appropriate for the maximum design temperature of the PRT.

Flying leads used to connect the head of the PRT to the resistance-measuring equipment must also be well insulated. It might also be necessary to consider dielectric leakage effects when an AC bridge is being used as, for example, PVC is a poor insulator at frequencies above 100 Hz. Older style PRT's had flying leads permanently attached to the internal lead within the thermometer's head structure and made use of *screw-down terminals* at the measuring instrument. However, the quality of plugs and sockets is now sufficiently good for there to be little additional electrical noise introduced into the measurement by their use. Moreover, it is much more convenient to handle a thermometer without a few metres of leads trailing from it.

Shielding is needed if long leads are used or if the environment is electrically noisy. Sometimes special precautions are needed to suppress unwanted signals that can enter the measuring system through the leads, for example, from TV or radar transmitters. Winding the thermometer leads through a co-axial high-frequency choke can sometimes assist.

Varying magnetic fields from, for example, motors can induce noise into PRT leads and it is very difficult to shield them from this kind of interference. Care should be exercised in the routeing of cables to minimise such noise.

2.7 IPRT sheaths

The traditional laboratory platinum-resistance thermometer has a Pyrex-glass or fused-quartz sheath or stem. These materials are very convenient, being easy to clean, and their transparency allows inspection of the inner components. However, they are very fragile and unsuitable for general industrial and much laboratory use (as the breakage rate of liquid-in-glass thermometers illustrates). Most IPRT's are therefore protected by metal sheaths (see Figure 2.3) and the type of material is set by the upper temperature limit. Most use a stainless steel, with Inconel[1] being the material chosen for the highest temperatures. The internal surface of the sheath must be clean and dry to prevent contamination of the sensor. Contamination of the Pt becomes a major problem at elevated temperature. It is also necessary for the sheath not to have cracks or holes, which may permit the ingress of contaminants or moisture. The most suspect part is around the sheath's sealed end, where the integrity of the weld can be at fault. Sheaths should be sufficiently long to

---

[1] reg. trademark of the INCO group of companies

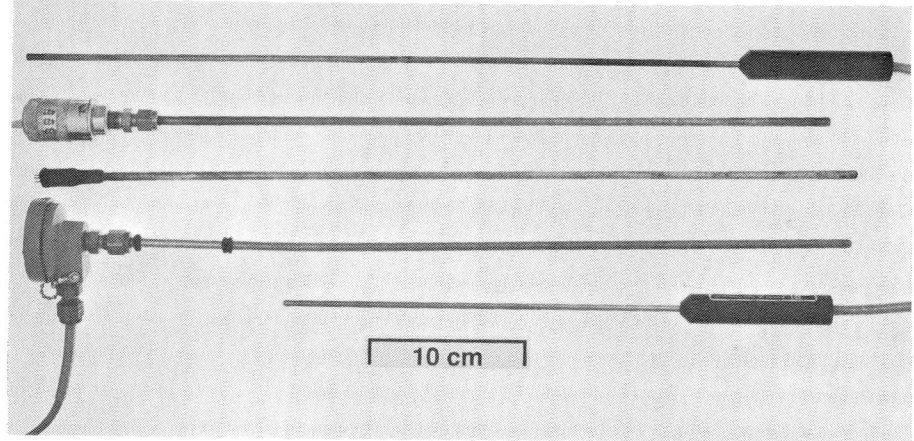

Figure 2.3: Typical industrial PRT's

allow adequate immersion of the thermometer in calibration or use.

2.8   Calibration

A *calibration* requires an approach that differs with the instrument. The 'calibration' of a resistance bridge consists of two parts (section 3.6.1): the determinations of (a) the corrections to be applied to each bridge step to make it 'linear' and (b) the correction to be applied to the bridge readings to convert them to absolute ohms.

Because of the method of scale 'pointing', liquid-in-glass thermometers can have discrete scale errors at the pointing marks, which causes the typical zig-zag shape of their correction curves. The calibration of these instruments involves determination of the errors at the pointing marks and checking the linearity between them. The bulb volume changes with time and the resultant error is corrected by measurements at a reference point, usually 0°C.

Thermocouples, on the other hand, are electrical instruments and have a smooth characteristic. They are normally calibrated [26] by measurements of the differences between their emf's and the corresponding values of reference functions, at a number of suitable temperatures. From the data, deviation functions are produced, either analytically or graphically, to allow interpolation. The choice of calibration points is not dictated by the instrument, as is the case with liquid-in-glass thermometers.

A PRT is normally *calibrated* by determining the coefficients of a polynomial that relates W to temperature—an interpolation equation chosen as appropriate to the temperature range and precision required.

IPRT calibration

In Chapter 1 the methods used to establish the ITS–90 were discussed, at least for the range from about −40 to 962°C. Although the SPRT interpolation thermometers must use highly-purified Pt wire, experience with previous temperature scales would suggest that the Pt in any *wire* IPRT sensor follows similar resistance-versus-temperature relationships. However, ITS–90 is set up differently to previous scales by using a reference function and individual deviation functions as the means of interpolation.

IPRT calibrations are often presented in the form of the interpolation equations defined in ITS-90 even though, in general, these PRT's do not meet some of the Pt-wire-purity specifications of ITS-90. However, it is generally accepted that uncertainties arising from this cause are not of any significance when the stability considerations are taken into account. Having said that, the ITS-90 interpolation equations require some complicated arithmetic to extract temperature and there is no need to use them for IPRT's. In fact, any calibration procedure and interpolation scheme that can be related to ITS–90 with the required uncertainty will do.

The form and complexity of the calibration procedure will depend on both the accuracy and the temperature range required. Platinum has a reasonably linear *R*-vs-*T* relationship. A linear calibration for the interval 0 to 100°C results in a maximum error of less than 0.4°C at 50°C, so that for narrow ranges a two-point calibration is all that is necessary.

Even if a quadratic interpolation function (equation 2.2):

$$R_t = R_0 \left[1 + At + Bt^2\right]$$

is considered, the value of the constant B ranges from about -5.75×10^{-7} to -5.89×10^{-7} for the different purity levels of the Pt wire used in PRT sensors. If a mean value of B is assumed the resultant uncertainty in a temperature measurement at 50°C would be only about 0.01°C. Thus, if a third calibration point is used to determine the value of B, the determination of the temperature of that calibration point has to be known to better than 0.01°C to significantly reduce the uncertainty in B relative to that of the mean value, mentioned above.

However, as the temperature range, and to a lesser extent the precision required, increases, three or more calibration points become necessary to determine the coefficients of the required polynomial, representing the PRT's resistance-versus-temperature relationship. As stated above, this discussion is considering IPRT's calibrated to working uncertainties better than about 0.05°C, at least in the lower temperature region. Thus, an interpolation equation must follow ITS–90 temperatures to within a few mK to avoid building into the calibration a significant systematic error.

In the case of the now obsolete IPTS–68, the interpolation formulae that applied to an SPRT consisted of a thermometer-dependent, quadratic function expressing its resistance ratio as a function of temperature, together with a thermometer-independent correction equation. IPRT calibrations were normally presented in a similar form.

When ITS–90 was introduced various forms of interpolation equation were examined to assess their closeness of approximation to ITS-90, for the various temperature ranges and uncertainties needed by IPRT's. It was found that the simplest way to present the interpolation equations was as polynomials of sufficient degree to meet the precision requirement mentioned above. The interpolation equations used by the National Measurement Laboratory (NML) are outlined in section (2.9).

User calibrations

Few laboratories have the need or the resources to set up fixed-point standards. So calibration by intercomparison in a heated enclosure is the more common approach. Obviously, the spatial and temporal variations of the enclosure have to be small enough to satisfy the accuracy requirements of the calibration, and need to be checked from time to time. The general procedures for such checking are discussed in section 8.4, but were oriented towards liquid-in-glass thermometer calibrations. While it is desirable to calibrate liquid-in-glass thermometers as the temperature is rising, to avoid 'stiction', a stable temperature is better for IPRT's, to avoid errors arising from differences in time constant. It is quite common to incorporate a metal block in the enclosure to smooth out controller fluctuations.

The continuous nature of the resistance-temperature relationship for PRT's results in there being no need for any set calibration temperatures. The only requirement is that they cover the required range. Extrapolation outside the calibration range is possible and, in fact, IPTS–68 required it for SPRT's between 419.58 and 630.74°C. However, problems with non-ideal behaviour at high temperatures make it a dangerous practice in the case of IPRT's. If it is necessary to extrapolate the calibration much above the maximum calibration temperature, at least the insulation tests described in section 2.11 should be carried out at the maximum extrapolated temperature, to ensure that there is no significant shunting of the sensor by imperfect insulators.

Because the resistance ratio (R_t/R_0) is used as the dependent variable in all PRT calibrations it is normal procedure to use the ice point as one of the calibration points.

2.9 IPRT calibrations at NML

Both fixed-point and intercomparison techniques are used at NML, depending on the temperature range and uncertainty required. The fixed points used are those set up for SPRT calibrations, and several IPRT's can be calibrated sequentially in the same 'point' by using each in a slow-melting situation. Because of the various constraints on the use of fixed points for IPRT calibrations, for example, dimensional requirements, the bulk of IPRT calibrations are carried out by intercomparison with SPRT's in a stirred-oil bath or other heated enclosures.

2.9.1 −40 to 250°C

For this temperature range it has been the practice in this laboratory to calibrate thermometers by intercomparison with an SPRT in an automatically-controlled stirred-liquid bath. The fluid used can be silicone oil, water or alcohol, depending on the temperature range. The spatial and temporal uniformity achieved in the NML stirred-liquid baths is 2 to 3 mK. The intercomparison measurements are carried out at a sufficient number (normally 14) of temperatures and a least-squares routine used to obtain the best-fit equation. With the introduction of the ITS–90, data from many earlier calibrations on IPTS–68 were reprocessed to refer to ITS–90 and it was found that a cubic equation,

$$W_t = 1 + A\,\frac{t}{100} + B\left[\frac{t}{100}\right]^2 + C\left[\frac{t}{100}\right]^3, \qquad (2.6)$$

fitted to about the same precision as was obtained previously on IPTS–68. A cubic equation in this form is not the most convenient way to extract temperatures from resistance-ratio values and instead it is usual to use the form:

$$t = 100\,\frac{[W-1]}{A} - 100\,\frac{B}{A}\left[\frac{t}{100}\right]^2 - 100\,\frac{C}{A}\left[\frac{t}{100}\right]^3, \qquad (2.7)$$

with a looping, successive approximation program to solve for temperature. The other approach is to fit the data by least-squares to temperature as a function of resistance ratio:

$$t = a\,[W-1] + b\,[W-1]^2 + c\,[W-1]^3. \qquad (2.8)$$

This gives a similar fit and is much more convenient to use.

The fits for two typical IPRT's are given in Figure 2.4. The difference between the goodness of fit for these two thermometers may reflect differences

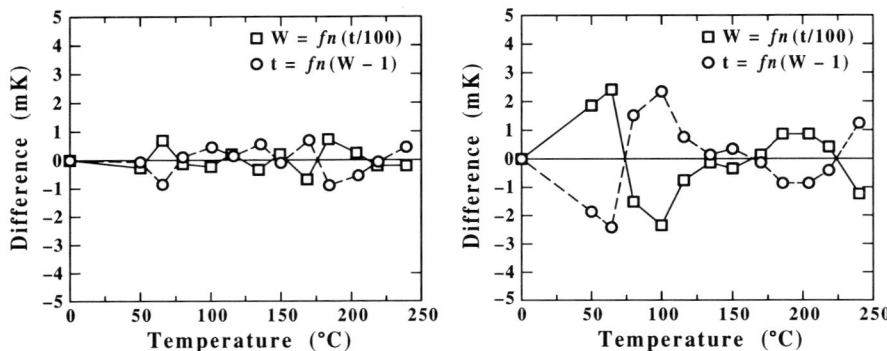

Figure 2.4: Graphs showing typical scatter of measured data from the least-squares fits using equations (2.6) and (2.8).

in the internal structure of their respective sensors, but since the difference is no greater than the estimated measurement uncertainty of 0.005 °C no definite conclusion can be drawn.

2.9.2   0 to 420 °C

It is not practical to operate the stirred-oil baths at NML at temperatures much above 250°C, so some calibrations are carried out using fixed points whose temperatures are defined on ITS–90. It is assumed that all wire PRT's follow the ITS–90 interpolation equations, at least to the uncertainty required by the users of IPRT's. The data from Zn and Sn freezing or melting points are used to determine the coefficients of the deviation equation (1.8). The uncertainty of calibration is estimated with the help of measurements at extra points, such as the freezing point of Cd and the melting point of Ga.

The use of the ITS-90 primary fixed points for IPRT calibrations is not always desirable because of size constraints and also possible damage to the glass thermometer wells used in these cells. Then, PRT's are calibrated up to 240°C in an oil bath, as described in section 2.9.1, and above 380°C in a caesium gas-pressure-controlled heat-pipe [27] (see also page 154). Interpolation from 240 and 380°C presents no problems because the resistance-versus-temperature relationship is adequately defined by the data obtained outside the interval.

The ITS–90 form of interpolation equation was judged as being inappropriate for the needs of most users of IPRT's, so the calibration is presented as a polynomial equation generated from the calibration data.

If fixed points are used to evaluate the coefficients of the relevant ITS-90 deviation functions, these functions in combination with the ITS-90 reference

function are used to generate resistance-temperature data in equal temperature steps over the range of calibration. These data are then put through a
least-squares fitting routine and the best-fit polynomial, consistent with the
uncertainty of the calibration, is used as the interpolation equation.

On the other hand, data obtained by intercomparison in the oil bath and
caesium heat pipe are used directly in a least-squares-fitting procedure to
generate the polynomial coefficients.

For the range 0 to 420°C, a fourth-order polynomial was found to be
necessary to keep systematic deviations from ITS-90 below 3 mK:

$$W_t = 1 + A\left[\frac{t}{100}\right] + B\left[\frac{t}{100}\right]^2 + C\left[\frac{t}{100}\right]^3 + D\left[\frac{t}{100}\right]^4. \qquad (2.9)$$

Since this equation is not the most suitable for obtaining temperature from
a measured resistance, the data are also fitted to the reverse function:

$$t = a\left[W - 1\right] + b\left[W - 1\right]^2 + c\left[W - 1\right]^3 + d\left[W - 1\right]^4. \qquad (2.10)$$

2.9.3   Calibrations above 420 °C

The experience of this laboratory has been that the stability of most IPRT's
deteriorates rapidly when used at temperatures above 460°C and that there
are few such situations requiring a better accuracy than can be obtained with
thermocouples. However, as more industries become energy-usage conscious,
the situation is changing and and there is an increasing requirement for
improved uncertainty for temperatures in the region from 500 to 750°C.
One area that does require relatively high precision at about 550°C is the
measurement of steam temperatures in power-station turbines.

Calibrations to 550°C or higher are carried out by first establishing that
the IPRT can survive about 16 hours at the highest required temperature.
Intercomparison calibrations can be carried out over the temperature range
from 0 to 1000°C using the oil bath and pressure-controlled caesium and
sodium heat pipes. Heat pipes, described on page 154, can provide a stable
furnace temperature uniform to within a few mK over some tens of centimetres.
The temperature is set by establishing a fixed gas pressure at the heat pipe's
condenser.

The form of interpolation equations applicable to such calibrations is still
being investigated, although equations similar to equations (2.9) and (2.10)
seem satisfactory with deviations from ITS-90 of less than 0.010°C.

2.9.4   Recalibration

The question of recalibration periods for IPRT's is an important one. As mentioned above, small changes in R_0 are not generally reflected in significant changes to the calibration coefficients, since interpolation equations are given in terms of resistance ratios. NML calibration certificates usually specify the allowable change in R_0 before the calibration becomes invalid. This assumes that the *current* value of R_0 as measured on the user's system is used to derive the temperature.

2.9.5   Summary

The most practical method of calibrating IPRT's is by intercomparison with an SPRT in a constant-temperature enclosure at a number of temperatures. For low-temperature calibrations (to 250°C) stirred-liquid baths can be used, and fluidised-bed baths and heat-pipe furnaces are more suitable at higher temperatures. The constants of the calibration equation are either obtained by least-squares fitting the measured data to an appropriate polynomial equation or, if a computer is not available, by choosing the ice-point resistance and a sufficient number of other spaced points to solve for the coefficients directly. Any extra points should be used to check the calibration.

2.10   Uncertainty of calibration

The least-squares-polynomial procedures, described above, yield coefficients that represent the calibration data, and the assignment of uncertainties to these coefficients must take into account a number of considerations. The RMS deviation of the experimental points from the fitted curve would be one measure, as would be the ice point (R_0) stability of the thermometer during initial heat treatment and also during the calibration. This assumes that other systematic errors are small or have been reduced to insignificance by correction procedures.

In calculating the uncertainty of calibration for IPRT's, NML makes some allowance for their expected behaviour under ideal conditions, of which we have some historical knowledge. The estimate takes into account the overall performance of the PRT during the calibration tests as well as the behaviour of similar types calibrated previously. The R_0 stability is then used as the criterion for recalibration. However, no allowance is made for errors arising from the user's techniques, measurement location or resistance-measuring instrument as these can be determined only by the user (see below).

Typically, the calibration uncertainty for a good-quality IPRT for the temperature range 0 to 250°C would be 0.02°C, and calibration validity would be maintained for at least five years with careful handling. This value is an expanded uncertainty [21] determined at a confidence level of 95%.

2.11 Errors in measurement

When using a calibrated PRT an assessment of the uncertainty involved must take into account potential errors from a variety of sources. One that is always present is that covered by the uncertainty of calibration. Others include that due to the conduction of heat along the sheath and lead wires, the heating effect of the measuring current, response time differences, non-ideal electrical insulation and electrical interference affecting the resistance-measuring equipment. This section deals with some of the more common systematic errors that can influence an IPRT temperature measurement—the methods of combining and expressing their effect are dealt with elsewhere [21, 22].

Current-heating effect

As mentioned in several places above, the measuring current will dissipate power in the sensor, raising its temperature above that of its surroundings. The magnitude of the rise depends on the thermal resistance between the sensor wire and the environment whose temperature is being measured. When the thermometer is in intimate contact with the object/medium of interest, it is the thermal resistance defined by the PRT's structure that matters, and a change in power (measuring current) is the only means available of altering the heating effect. The rise in temperature, Δt, resulting from a measuring current, I, is proportional to the dissipated power:

$$\Delta t \propto I^2 R. \tag{2.11}$$

This relation illustrates two points:

1. the resistance of the PRT, (its R_0) should be kept as small as possible while being consistent with the sensitivity required,

2. the measurement current should be kept to a minimum.

As an example, the rise in temperature of a typical IPRT, immersed in an ice point, with a 1 mA current will be about 0.01°C. In a similar situation a current of 10 mA will cause a temperature rise of nearly 1°C—highly significant if an uncertainty of 0.05 °C is being sought. Equation (2.11) may be used to correct for the effect. If measurements are taken at two different currents the equation will enable extrapolation to the zero-current resistance. Higher class resistance-measuring instruments designed for use with PRT's often have a $\times \sqrt{2}$ current multiplier to simplify the extrapolation.

In most situations the magnitude of the heating effect is not a serious problem, provided appropriate resistance-measuring equipment is being used. Furthermore, interpolation formulae always incorporate resistance as a ratio and, as the effect is proportional to the resistance, use of resistance ratio tends to minimise the systematic error. Manufacturers' specifications often provide data on the heating effect in their product. A word of warning: some digital multimeters use a measurement current of 10 mA for the resistance range used with IPRT's. As pointed out above, this current is too large for 100 Ω sensors.

Immersion errors

Thermal conduction from the sensor, along the sheath and leads to the outside environment, will prevent the thermometer attaining the temperature intended. It is influenced by the distance the thermometer is immersed into the region being measured, by the type of material in the sheath, the leads and their insulation and by the sheath diameter. A 6 mm diameter IPRT with a stainless-steel sheath and alumina four-bore insulation requires about 150 mm immersion into a normal ice point to reduce any possible error to below 5 mK. The test for adequate immersion is to withdraw the PRT by a small amount and observing if there is any change.

Response-time effects

The time taken for a thermometer to respond to a step change in temperature is of importance in considering the overall uncertainty in a temperature measurement. The parameter involved is known as the *time constant* and is defined as the time taken for the temperature of a thermometer to change by $1/(1 - e)$—about 68%—of the applied temperature excursion. The magnitude of the time constant is largely determined by the internal structure of the thermometer and its sheathing, but there are some effects due to the principal mode of heat transfer predominating in the particular temperature region being considered. Thermal radiative-heat transfer is rapid, whereas conduction is controlled by the thermal diffusivity of the materials involved and can be quite slow. If the temperature of the environment being measured is changing at a linear rate, the thermometer will lag behind and will indicate a temperature of the environment at a time corresponding to the *time constant* earlier.

Thus, the time constant is an indication of the error due to thermal lag in the system—for a linear temperature rise the error is the negative of the rate of change of temperature times the time constant. This can be important when calibrating thermometers by intercomparison. Either the standard and test thermometers should have similar time constants or the rate of temperature change of the heated enclosure being used should be kept relatively small. In a bath with a sinusoidal control cycle it is possible to have one thermometer indicating a rise in temperature while another registers a fall.

Electrical-leakage effects

Electrical leakage is a major source of error in PRT's at higher temperatures. Leaky insulators shunt the sensor and will cause deviations from the expected interpolation equations. Shunting of a $100\,\Omega$ (R_0) sensor by a leakage resistance of $1\,M\Omega$, a not uncommon situation for a PRT, will cause an error of about $60\,mK$ at about $300°C$. The situation becomes dramatically worse at temperature above $700°C$. IEC–751 requires all PRT's to have a measured resistance between the thermometer leads and the sheath at room temperature of more than $100\,M\Omega$, and type tests specify $> 10\,M\Omega$ from 100 to $300°C$ and $> 2\,M\Omega$ from 300 to $500°C$. In some situations these criteria are still insufficient and a better insulation resistance can be necessary.

The effect, if present during calibration, will be included in the interpolation equation so determined. However, this will not apply to any extrapolation region where no actual measurements were obtained. Hence the earlier warning on the danger of extrapolation outside the calibration range.

It is possible to obtain a measure of possible insulation resistance by measuring the resistance between the PRT's sheath and its leads with a megohmmeter. This will give similar values to inter-lead shunting. A low-voltage megohmmeter is to be preferred as high-voltage instruments may cause arcing and permanent damage.

It should be noted that, while shunting will reduce the apparent resistance of the sensor a short-circuit between a current and a potential lead that are connected to the same side of the sensor will cause a rise in the measured resistance. This is because the branch point is moved from the sensor's termination, adding in extra resistance.

Lead resistance

In most modern resistance-measuring equipment a large lead resistance does not cause appreciable error, apart from increasing the possibility of inter-lead leakage or, if due to a long-lead length, electrical noise. However, there are some instruments, particularly some AC bridges, where the inter-lead capacitance associated with long leads can result in the circuitry oscillating and saturating the input circuits. The manufacturer's specifications will provide appropriate guidelines.

It should also be remembered that low internal-insulation resistance not only shunts the thermometer sensor, but can also cause serious error, by providing a leakage path through the thermometer surroundings to earth, thereby shunting the resistance bridge or possibly damaging electronic circuits.

Vibration

Excessive vibration and shock can work-harden the Pt wire in the sensor causing a rise in R_0. This will be relieved by annealing at temperatures above about 430°C. Severe shock can cause permanent changes to the sensor, necessitating recalibration. The thermometer's calibration certificate contains information on allowable changes to R_0 that can occur before recalibration is necessary.

Summary

There are other sources of error, such as electrical interference, that can affect IPRT's, but generally these manifest themselves, at least to some degree, in noise and a discerning operator will be aware that there may be a problem.

The effect of the identifiable sources of error must be assessed and the uncertainty components arising from these combined with the calibration uncertainty to obtain the overall uncertainty, the 'expanded uncertainty' [21]. Methods for combining such estimates are treated in references [21, 22].

2.12   IPRT selection

Very few people are likely to construct their own resistance thermometers, although it is desirable to know the principles involved and the design compromises that must be made. There is a wide variety of commercial thermometers available and without adequate background it is impossible to make an informed selection or to decide which of the parameters in the standard-code specifications are important for a particular application. It is difficult to recognise sources of error and to ensure that they have been kept under control. The selection of an appropriate IPRT will obviously depend on the use being contemplated, so one must consider suitability of the sensor type, the lead insulation and the sheathing material at the expected temperature extremes. Physical dimensions, such as diameter and length, may be important as can the ability to withstand shock. However, a point that is often overlooked is the need for calibration and subsequent ice-point checks. The thermometer must be compatible with the equipment of the calibration laboratory, so it should be consulted before the final choice of thermometer is made. The necessity for periodic checks of R_0 has been emphasised as the best method of maintaining confidence in the thermometer calibration, but heat conduction down a short stubby sheath will prevent the sensor coming into equilibrium with the ice bath and will invalidate such checks.

When choosing an IPRT, consideration must be given to many aspects of the measurement. A few are listed:

1.	Maximum temperature	:	Can the internal insulation withstand it? Will the lead joints break down?
2.	Minimum temperature	:	Some ceramics will shatter if cooled too rapidly.
3.	Sheath material	:	Maximum temperature? Compatibility with the measurement environment?
4.	Thermometer length	:	Conduction errors? Can it be calibrated?
5.	Thermometer diameter	:	Conduction errors? Will it fit into the apparatus?
6.	Shock resistance	:	Check manufacturer's specification.

Chapter 3

Resistance Thermometer Measurement

John Connolly

3.1   Measurement of resistance

The use of resistance thermometers to measure temperature necessarily requires the measurement of electrical resistance, often at the most precise level. Before looking in detail at resistance-measuring equipment designed or suitable for use with PRT's, it is worthwhile beginning with a discussion of the general principles of resistance measurement.

A common device for measuring resistance is the ohmmeter, in a simple multimeter. A fixed voltage is applied to the resistor and the current is measured in an ammeter mode:

$$R = \frac{V}{I}. \tag{3.1}$$

The resistance is inversely proportional to the current and the ammeter will have an inverse scale, calibrated in ohms. This is the reason for the non-linear scale on an analogue ohmmeter, particularly at the high-resistance (low current) end. The ohmmeter is only useful over about two thirds of its scale and then only to 5 or 10%, but it is relatively easy to obtain a constant voltage source—even a dry-cell battery is good enough.

Some modern instruments, especially digital multimeters, use a constant current source and measure the voltage drop across the resistor to be measured. Referring to equation (3.1), it is seen that the resistance is now proportional to the voltage (fixed I) and the voltmeter, when scaled in resistance, will be linear. The resolution on an analogue meter can be as good to 0.1% and, in

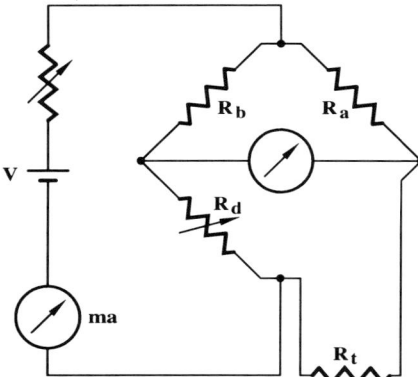

Figure 3.1: Circuit of a Wheatstone bridge showing the bridge resistors, R_a, R_b and R_d, and the resistance R_t being measured. V is a constant voltage source that supplies current via two connections to the resistance bridge and a galvanometer is connected to the other two.

the case of digital instruments, improved resolution results in an increase in the number of digits displayed and thus the cost. The accuracy is usually limited by other considerations.

Wheatstone bridge

The Wheatstone bridge (Figure 3.1) is a simple network of resistors (or other impedances) and can be used to determine the value of an unknown resistor, R_t, in terms of an adjustable reference resistor, R_d, and the ratio of two others, R_a and R_b. At balance no current flows through the galvanometer and the potential difference across it is thus zero. Hence:

$$\frac{R_a}{R_b} = \frac{R_t}{R_d}$$

$$\text{or} \qquad R_t = R_d \, \frac{R_a}{R_b}. \qquad\qquad (3.2)$$

R_d is normally a calibrated decade resistor and R_a and R_b are referred to as the ratio resistors. The Wheatstone bridge, at balance, is independent of the supply voltage, variations of which alter only the galvanometer sensitivity. By using ratio resistors whose ratio differs from unity, a single decade box can cover many ranges. It is also possible with an unequal ratio to measure low-resistance values where switch-contact resistance in an equal-ratio bridge

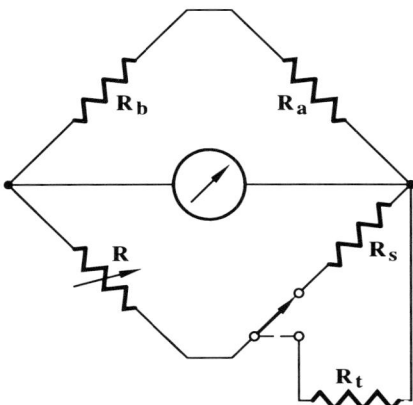

Figure 3.2: A variation of the bridge in Figure 3.1, in which resistor R is set to balance a sensor R_t at a specific temperature. A single-point calibration is possible by switching in a reference resistor R_S.

would invalidate such measurements. However, other error sources, such as power-dissipation effects and finite lead resistances must be taken into account.

Resistance thermometers are widely used as sensors in a variety of direct-reading sensor/indicator combinations. Typical examples include the monitoring of motor vehicle coolant temperature, air-conditioning and refrigeration. The output of resistance-measuring instruments is also used for temperature control.

If resistance R_t in Figure 3.1 represents a resistance thermometer the resistance of its leads would be in series and will be included in the measurement. As the leads will not, in general, be at the same temperature as the sensor nor will they have the same temperature coefficient of resistance, their resistance will limit the accuracy obtainable—1% is about the best uncertainty that can be expected from such an instrument, except in the case of some high-resistance thermistors, mentioned in section 2.4.

In some instruments the adjustable decade resistor, R_d, is calibrated in terms of temperature.

Another variation of the Wheatstone bridge, for use with resistance thermometers, is shown in Figure 3.2. If the adjustable resistor R is used to balance the bridge at a particular temperature (e.g., at 0 or 20°C) then the out-of-balance signal is almost linear in temperature, as long as the departure from the set value is not too large. Many cheap indicator instruments operate in this way. The limitations are non-linearity and the dependence of signal on bridge current (or voltage). Simple standardisation can be provided by a reference resistor, R_s,that can be substituted for the thermometer and the bridge supply altered until the output signal is at a set value. In such instruments the accuracy depends critically on the stability of the supply voltage and in car temperature gauges, for example, the line voltage can change erratically, and good voltage regulation must be introduced.

The other approach is to use a special detector, a 'ratiometer' that has two opposing windings whose combined effect cancels variations in the supply voltage.

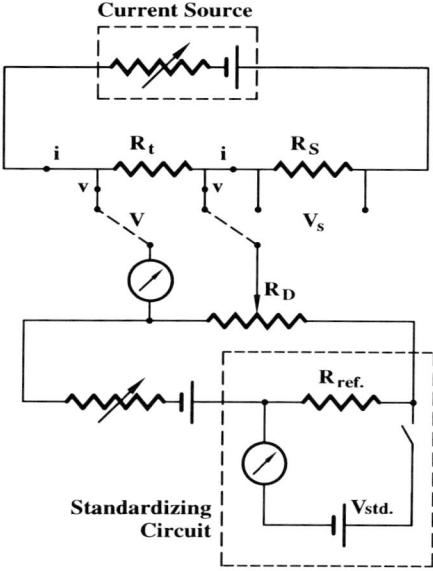

Figure 3.3: Circuit illustrating the potentiometric method of measuring a resistor R_t. The potential difference V across R_t and that across the reference resistor R_S, V_S, are measured by the 'potentiometer', indicated by the circuit below the switch.

Potentiometric method

The potentiometric method of measuring the resistance of a resistor R_t is illustrated in Figure 3.3, in which a constant-current source supplies current to a reference resistor R_s in series with R_t. The potential difference measured across R_s gives the current, $I = V_s/R_s$, and that across the resistance being measured, V_t, yields the result:

$$R_t = \frac{V_t}{I} = \frac{V_t}{V_s} R_s \,.$$

The method derives its name from the use of a 'potentiometer' (see Figure 3.3), a device that measures potential difference by comparing it with a known fraction (resistive divider) of a reference voltage, typically a standard cell. In practice, the current through the potentiometer is first 'standardised', i.e., adjusted until the potential difference across a reference resistor, R_{ref}, equals that of the standard cell, V_{std}.

The potentiometric method works well because the input impedance of the potentiometer (or the modern equivalent, the DMM, see section 3.4) approaches infinity at balance, hence no current passes along the potential leads to the resistor being measured. Thus, the measured resistance is that between the branching of current and potential leads and does not include a contribution from lead resistance. Sources of parasitic emf's, mainly in the potential circuit, will cause a bias to the voltage reading that can be removed by making a second reading with the current reversed and taking the mean. An alternate method is to take a reading with the current turned off and to subtract this from the primary reading. This will correct for a parasitic emf only if it is current-independent, not necessarily so for chemical-emf sources.

3.2 IPRT and SPRT measurement

An SPRT designed for use in the range -200 to $420\,°\text{C}$ normally has $R_0 \sim 25.47\,\Omega$, chosen to give the convenient temperature sensitivity of about 0.1 $\Omega/°\text{C}$. In order to limit the effects of current heating to a reasonable figure, a measurement current of 1.0 mA is usual. Consequently, for the potentiometric method, a DVM will require a range of 0.1 V and a resolution of $0.1\,\mu\text{V/mK}$, i.e., it needs to cover at least six decades. On the other hand, a resistance bridge designed for the same measurement will need a range of $100\,\Omega$ and a resolution of about $10\,\mu\Omega$. For other values of R_0 the measurement requirements will be scaled accordingly.

It is common to choose $R_0 = 100\,\Omega$ for an IPRT, which will require a resistance bridge with a range to, perhaps, $500\,\Omega$ and a least count of $1\,\text{m}\Omega$ or less. If a 1 mA current is used, a potentiometric device will need a range of up to 1 V and a least count (resolution) of $1\,\mu\text{V}$. Thus, even for an IPRT, the measurement precision required is often much greater than for most other resistance measurements.

Measurements with simple resistance-measuring equipment usually include the resistance of the sensor leads. On the other hand, instruments designed specifically for PRT's usually includes a method of lead compensation. The following sections cover some of these methods and extend to instruments that allow measurement to the highest precision.

The leads of four-lead thermometers are commonly marked with the letters 'c', 'C', 't' and 'T', and traditionally the C and T leads are connected as the potential leads. In practice, however, with modern instruments designed for use with PRT's, it does not matter which of the leads are selected to carry the current and which are to serve for the measurement of potential difference.

3.3   DC-bridge methods

Resistor stability

All the bridges described below use resistors as the principal circuit components, and resistors have characteristics that necessitate periodic calibrations and close temperature control. The resistance of each coil depends on the length of wire cut off by the maker, so an initial calibration must be carried out to determine the exact value of each step of the dial. Also, the wire is usually an alloy and its resistivity will change with both time and temperature. The time drift is probably due to slow recovery from changes induced by deformation during winding and, as one would expect, the drift normally decreases with time. The temperature-dependent variations arise from the finite temperature coefficient of resistance of the material used. More recent instruments tend to use Evanohm, as its temperature coefficient can be tailored to be close to zero at ambient temperature. For these reasons resistor-based instruments have to be calibrated at regular intervals as well as maintained and operated under well defined temperature (and humidity) conditions.

When the instrument is to be used for resistance thermometry the calibration checks involve determination of only the relative values of the various decade positions. This is because the interpolation formulae employed for the resistance-temperature relationships of PRT's are always in terms of resistance ratios and, hence, are independent of the absolute ohm values. Of course, this presupposes that both resistance measurements, used to determine the ratio, were measured on the same instrument under similar ambient conditions.

The recalibration problem can be reduced by the use of AC ratio transformers as bridge components, but then consideration has to be given to AC effects in the bridge and in the thermometer.

Instrument temperature

The resistors in a bridge or potentiometer produce thermal emf's when connected by copper wires if a temperature gradient is present. Changing temperatures produce drifts in residual emf's, which appear as drifts in the DC zero. Constantan cannot be used in resistors for high-quality work because of its large thermal emf relative to copper, so Manganin or Evanohm is preferred. A heavy instrument case is used to smooth out temperature differences. The resistors have a small but finite temperature coefficient of resistance, in old equipment often amounting to 20 ppm/°C at ambient temperatures and in modern instruments, mostly below 5 ppm/°C.

For platinum- or copper-resistance thermometers, a change of 4 ppm in resistance is equivalent to 1 mK and, if an uncertainty of a few mK is required the temperature of the higher-valued resistors in the decade of the instrument must be monitored and corrections applied if necessary. A simple mercury-in-glass thermometer is usually adequate for this. In some designs the resistors are mounted on the interior of the switch contacts, and the heat from the operator's hand can produce detectable changes after periods of 10 minutes or so of steady use.

This effect can be minimised by siting measurement equipment in an air-conditioned space or by placing critical resistors in temperature-controlled enclosures. For measurements requiring the highest precision the resistors are kept at a temperature where the temperature coefficient of resistance is close to zero—for Manganin this is at about 35°C. The other technique is to shunt the resistor with a compensating temperature-sensitive resistor. A 10 Ω Manganin resistor at about 20°C requires a few kΩ of copper in parallel with it. However, as mentioned previously, the use of Evanohm and similar low temperature-coefficient alloys have superseded such techniques.

With potentiometers, the relative value of resistors is the main consideration, and temperature control is not normally used. However, when comparing a thermometer with a standard resistor, the temperature of this standard is important and the comments made above on resistors in a bridge also apply.

3.3.1 Three-lead bridges (Siemens type)

The prime consideration in the design of Wheatstone-style bridges for PRT use is removal of the effects of lead resistance. One way is to increase the resistance of the sensor, so reducing the relative effect of the leads, but this gives rise to power dissipation effects and an increase in insulation-resistance requirements. Alternatively, a system of lead compensation can be used. In its simplest form only three of a thermometer's leads are connected, with one, the c lead, connected to one side of the sensor, and used as a current lead, as illustrated in Figure 3.4. This places the other two leads, C and T, in opposing arms of the bridge, and as it requires only three connections it is suitable for 3-lead thermometers. If the leads are adjusted to equal resistance, approximately, and an equal-ratio bridge is used, the value of R_x is equal to that of the element alone when the bridge is balanced.

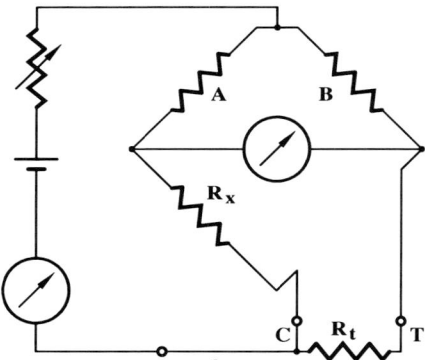

Figure 3.4: Siemens compensated-lead bridge, where A and B are the ratio resistors, normally equal, and R_t is the resistance of the thermometer sensor. Only three thermometer leads are shown: c, C and T.

This system is used in the resistance thermometer versions of many commercial controllers and chart recorders as well as in bridge instruments. It reduces, but does not remove, lead effects since the leads are never of exactly equal resistance. However, it does enable long leads to be used in factory installations. An equal-ratio bridge should be used for this configuration, because any lead resistance in series with R_x will be multiplied by the ratio and if the ratio is not equal to one the resistance of the lead, in the arm of the bridge, cannot be compensate for.

3.3.2   Mueller bridge

The Mueller bridge [28] is a variation of the Wheatstone bridge following from original work by Smith [29]. It is designed for precise measurements of the resistance of PRT's and has undergone considerable refinement, through the collaboration of the National Institute of Standards and Technology, U.S.A. with Leeds and Northrup Company Inc. Other manufacturers have also, in the past, marketed versions of the instrument.

The circuit is basically a three-lead Wheatstone system and two readings are necessary. Referring to Figure 3.5, if C is the resistance of the C lead, etc, then, in the *Normal* or *N* position:

$$\frac{P}{Q} = \frac{R_n + C}{R_t + T}$$

The leads are then switched to the *Reverse* or *R* configuration shown in (b) of

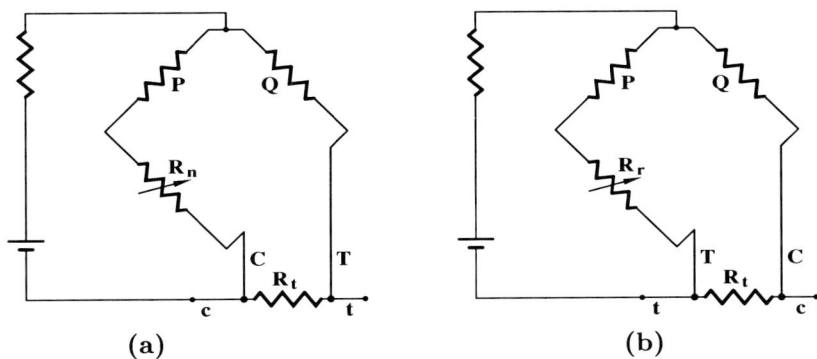

Figure 3.5: Mueller bridge. The diagrams show the positioning of the thermometer leads, c, C, t and T, in the circuit for each of the two measurements.

the figure:

$$\frac{P}{Q} = \frac{R_r + T}{R_t + C},$$

and since $P/Q = 1$:

$$R_t = R_n + C - T$$
$$\text{and} \quad R_t = R_r - C + T$$
$$\text{i.e.,} \quad R_t = \frac{R_n + R_r}{2}.$$

This result is independent of the lead resistance, provided:

- P = Q,
- T and C do not change during the measurement and
- the Normal–Reverse (N-R) switch-contact resistances are reproducible.

For high precision, circuits like Figure 3.6 are used. It incorporates a number of features, many of which are used in other bridge constructions:

(a) There is a slide-wire adjustment to equalise P and Q.

(b) Mercury contact switches, reproducible to $1\,\mu\Omega$ or better, are used for the N–R commutator.

(c) Switches for ($\times$ 1) and ($\times$ 0.1) Ω decades are in series with the resistors in the ratio arms, which are $500\,\Omega$ or greater, making variations in contact resistance unimportant.

(d) The small steps in resistance for the $(\times\ 0.01)\,\Omega$ and lower decades are achieved by the so-called *shunted-resistor* technique. If $1\,\Omega$ resistors are switched in parallel with a fixed $100\,\Omega$, the resistance of the combination is reduced in steps of approximately $0.01\,\Omega$. Thus, with a counter-balancing resistor in the opposite arm of the bridge (K in Figure 3.6), the lower-dial decade steps can be achieved, even down to $\mu\Omega$ values. The calculation of appropriate resistance values is an exercise in mental gymnastics.

(e) Mercury contact switches are used for the $(\times\ 10)\,\Omega$ decade to reduce contact resistance.

(f) A switch in the battery circuit (Figure 3.6) is used to reverse the bridge-supply current, as the mean position of the galvanometer for the *forward* and *reverse* current directions is a truer zero, free from parasitic emf's in the circuit, than is its position for an open circuit in the battery system. The current in the thermometer flows most of the time and variations in the current-heating effect are avoided.

(g) In some models, the commutator has 2 extra poles that reverse P and Q, so that on position 'N'

$$\frac{P}{Q} = \frac{R_n}{R_t}$$

and on 'R'

$$\frac{Q}{P} = \frac{R_r}{R_t}.$$

This reduces errors arising from the departure of the ratio P/Q from unity.

These, and indeed most, DC bridges are difficult to use, requiring two measurements for each resistance determination and much concentration and patience. Parasitic emf's in leads and elsewhere in the circuit cause shifts in the detector zero, making it difficult to determine when the bridge balance conditions have been reached. Drifts in lead resistances also cause noise in the measurement.

The Mueller bridge is a resistor-based instrument and must have its calibration checked periodically. Since the calibration of a resistor changes if subjected to shock or vibration, it is advisable for calibrations to be carried out *in-situ*, rather than in a test laboratory. The instrument is configured such that the decade steps are, effectively, in series, so a linear build-up calibration technique may be employed. A description of the procedures is given in section 3.6.1.

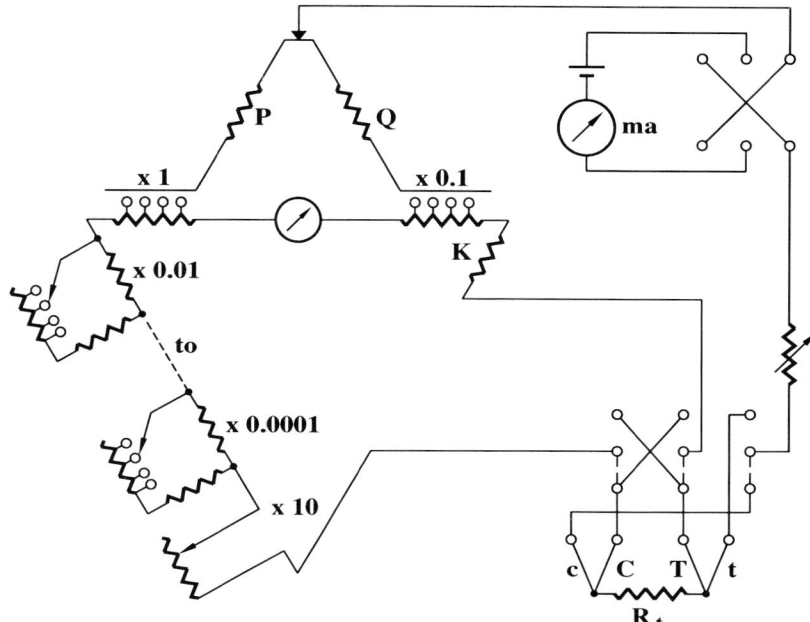

Figure 3.6: Shunted-dial precision Mueller bridge illustrating shunted resistors for the lower decades as well as the commutating switch arrangement.

Modern developments in instrumentation have made Mueller bridges obsolete and this description has been included only because there are several still in use and they are better than nothing.

3.3.3   Unequal-ratio bridges

The resolution (least-count) required of a resistance bridge for use with typical resistance thermometers was indicated in the introduction to section 3.2.

Consider the bridge of Figure 3.1 on page 56. The decade resistor R_d can be scaled by the ratio of R_a/R_b, as shown in equation (3.2). If, for example, $R_a/R_b = 0.01$, many of the resistors in the decade R_d can be of a more convenient size, because of the multiplying effect of the ratio, and those at the lower end of the decade may be made without the 'shunted-resistor' technique, mentioned in section 3.3.2. The resistor corresponding to a change of $1\,\mathrm{m}\Omega$ in the unknown would be $0.1\,\Omega$, which can be adjusted to better than a $\mathrm{m}\Omega$ ($\sim$ 0.1%) relatively easily. Hence, it is practicable to have an effective least count on the bridge of $10\,\mu\Omega$ or less. However, steps of $100\,\Omega$ in R_t then require steps of $10\,\mathrm{k}\Omega$ in the decade and such resistors must be very well constructed

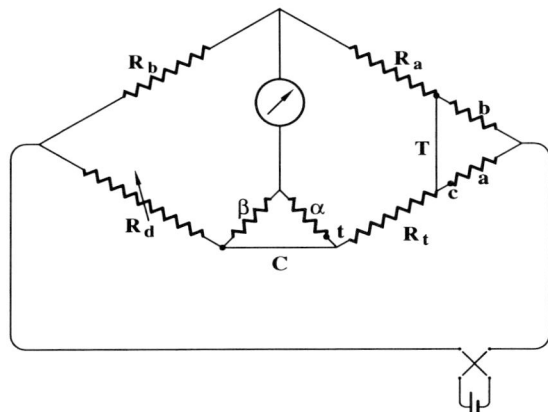

Figure 3.7: Kelvin double bridge places the thermometer leads, c, C, t and T, into the circuit so that the multiplying effect of the ratio resistors, R_a and R_b, can be used to advantage in the bridge circuitry, while not affecting the lead compensation.

to provide a stability that warrants the 0.1% adjustment in the 0.1 Ω resistors.

A major advantage of the design is that switch resistances need to have only 1% of the reproducibility needed in an equal-ratio bridge and, in cheap instruments, the possibility of using low-cost switches is important. However, special procedures or circuitry have to be employed to compensate for lead resistance.

3.3.4   Kelvin double bridge

The 'leads' in an unequal-ratio bridge are not balanced, if a simple three-lead design is used. However, a Kelvin double bridge can be employed using a four-lead configuration and a second ratio bridge shunted across the leads. Figure 3.7 is a schematic of the arrangement. The effect of lead resistance is removed by a separate adjustment, which is less critical than the main measurement and is likely to vary only slowly.

This bridge requires two balances for a temperature measurement, as does the Mueller bridge, although a suitable design of Kelvin bridge makes one balance much more important than the other, especially if the leads are approximately equal. Bridges of this type, described by Smith [29], are still marketed by British companies as laboratory instruments, but have now been superseded by AC ratio-transformer-based resistance bridges. (The Mueller bridge was also, in fact, a Smith design.)

3.4 Potentiometric methods

The principles of the potentiometric method of measuring resistance were dealt with in section 3.1. There are some complications when this method is used to measure PRT's because of the very small voltages being measured and the possibility of thermo-electric and other spurious emf's being generated in the PRT leads and in other components of the potential circuit. As potentiometers normally do not read negative voltages, the input must also be reversed when the thermometer current is reversed. The reversing switch should be chosen to not introduce significant spurious emf's.

However, the method is now very attractive for IPRT measurement, not because of potentiometers as such, but as a result of the availability of high-quality digital multimeters. These instruments have a very high input impedance and are, therefore, the modern equivalent of potentiometers. They can be used in either their ohm or millivolt modes, the latter being used in conjunction with an external constant-current supply. Digital voltmeters do not require input reversal on reversing the PRT current because, in most cases, they can measure negative voltages. However, some show a bias when the input is reversed and this will cause an error.

Lead compensation

Compensation for lead resistance in 3-lead thermometers is achieved using the bridges of section 3.3.1, but these require an assumption of equality for the resistances of two of the leads. The Mueller bridge (section 3.3.2) compensates for lead resistance in 4-lead thermometers, by making measurements with three of the leads twice, and requires that lead resistances remain constant during the measurements.

In a true 4-lead, or potentiometric, measurement two of the leads carry current to and from the sensor and the other two allow measurement of the potential difference across the sensor. In Figure 3.3 (page 58) the two lead wires marked 'i' supply the current to the PRT sensor and voltage is measured using leads 'v'. Since the voltage is measured with a high-impedance detector no significant current flows in the potential leads and the measured voltage is that developed across the sensor—between the two junctions of the i and v leads.

Ratio-transformer bridges, described in section 3.5, operate in a similar way and are better described as AC double potentiometers.

3.4.1   Digital ohmmeters

Most modern digital voltmeters (DVM) are, in fact, digital multimeters (DMM) and can be operated as ohmmeters. They include all the circuitry equivalent to that shown in Figure 3.3, where the DVM section replaces the potentiometer. In addition, better-class instruments have a measurement procedure for removing offsets caused by parasitic emf's in the voltage leads, usually by taking a measurement with the current switched off. However, DMM's do not have provision for altering the measuring current and care must be taken in selecting an instrument to ensure that the current used on the resistance range is compatible with the PRT and will not cause excessive current-heating (see page 49).

There are available a number of digital thermometers—integrated sensor/instrument combinations—that use PRT's as sensors and incorporate microprocessors for data processing and manipulation. Some have provision for the entry of the sensor's calibration coefficients and display directly in temperature with a resolution of 1 mK or better. The topic is discussed more fully in Chapter 4.

3.4.2   DVM-based systems

In cases where the current employed in the ohm range of a DMM is inappropriate, a separate current source can be employed and a system equivalent to that of Figure 3.3 may be used. However, such a system has an even greater capability if the current source is able to excite several PRT's and a circuit like that of Figure 3.8 is used.

One standard resistor can be used in series with several resistance thermometers and, as long as the current is stable, several thermometers can be measured in quick succession. The self-balancing feature of the digital voltmeter permits automatic data acquisition as well as automatic control of intercomparison baths, etc.

As there will be a variety of metals present in a typical circuit, the small voltages being measured are likely to be biased by emf's arising in the voltage circuit from thermal and chemical effects. If these vary only slowly with time and do not depend on current direction they can be removed by reversing the current and taking another set of readings of V_s and V_t. The mean values remove the bias. The accuracy of the system relies on the same current flowing in the PRT's and in the reference resistor. Low insulation resistance in IPRT's, between their leads and metal sheath, can result in significant leakage of the measurement current to earth, through surrounding metallic components to the low potential terminal of the current source. Thus, the current flowing in the reference resistor will be different from that in the PRT's. The effect

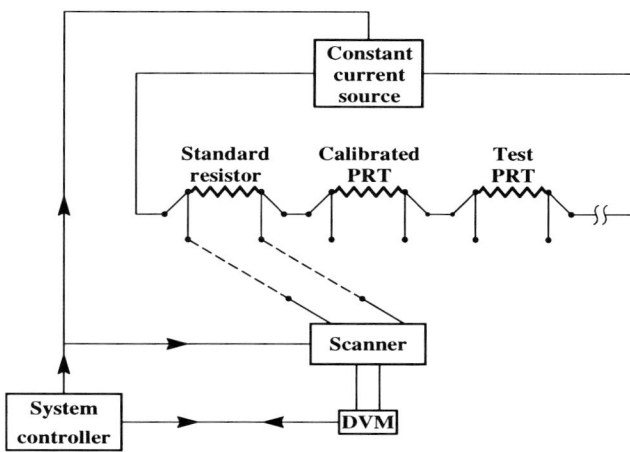

Figure 3.8: DVM based data-acquisition system. A four-lead configuration is required and if the branch points are close to the sensor, the lead resistance problem is removed.

is accentuated by current reversal because the effective potential relative to earth of the leaky thermometer will change as will the leakage current. This is a further reason for the requirement that the insulation resistance between the sheath and leads of an IPRT should be greater than 100 MΩ. Scanner switches, particularly those employing solid-state components, can have substantial leakage to earth, or to their power supply, in their *off* state and will cause similar errors.

3.5   AC-bridge methods

It is possible to use AC supplies with normal Wheatstone bridge circuits and with potentiometers. Then, the detector amplifiers are simpler and spurious DC signals are avoided. However, calibration and stability problems remain and provision has to be made for the balancing of reactive components in the circuit.

3.5.1   Ratio-transformer bridges

A ratio transformer is a multi-tapped transformer, wound in such a way that voltage ratios between the various tappings of the secondary windings depend only on the relative turns ratios. In other words, the transformer is made 'perfect' by ensuring that all turns are linked by all the magnetic field lines and

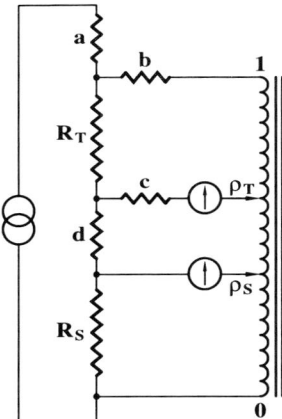

Figure 3.9: Use of a ratio transformer in an AC bridge circuit configured more like a potentiometer circuit than a traditional bridge. Resistor R_T represents the thermometer resistance

that non-linearities arising from non-ideal behaviour of the core are minimised. The techniques for accomplishing this are well established and transformers with uncertainties approaching 0.001 ppm have been constructed.

Instruments, such as that illustrated in Figure 3.9, use adjustable ratio transformers as the equivalent to the decade resistors of the instruments described in the previous section. They will not drift with time and will only need an initial linearity calibration to confirm their accuracy because the established ratios of voltage or current are set by the turns ratio alone.

Much of the circuit complexity is aimed at removing the effect of the resistance of the link, d, between R_S and R_T (Figure 3.9), the link being part of the lead configuration of a four-lead resistance thermometer. The other (potential) leads are in series with the high impedance of the transformer winding and their combined resistance is unimportant if it is not too large. There are a number of commercial bridges based on this principle, but they are fairly expensive laboratory instruments.

Figure 3.10 illustrates an AC bridge [30] developed at NML in the early 1970's. A further example is that of Figure 3.11, a schematic of one of the instruments produced by Automatic Systems Laboratories of Milton Keynes, U.K. who have specialised in such devices.

The carrier generator produces a pure sinusoidal constant current that flows in turn through the reference resistor R_S and the PRT resistance R_T, as illustrated in Figure 3.11. The voltages produced across each are in exact ratio to their resistances. The voltage V_s, across R_s, is applied to the primary

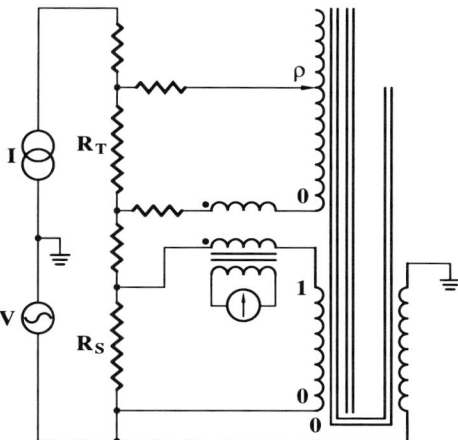

Figure 3.10: High-precision AC bridge—an NML design based on a ratio transformer. Shown is the extra field-excitation winding on the transformer that is crucial
in ensuring the precision of the linearity of the decade winding. At balance, R_T/R_S
$= \rho$, the transformer ratio.

winding of a high-precision ratio transformer. The voltage tapped from
the adjustable secondary winding is compared with the voltage across the
PRT, the difference being amplified by a low-noise, high-gain detector. The
transformer ratio, n, is adjusted, either manually or under automatic control
of a microprocessor, until the output is zero. Whence:

$$\frac{R_T}{R_S} = \frac{V_T}{V_S} = n. \tag{3.3}$$

There is also an automatic quadrature balance to remove reactive components from the balance. The effects of lead resistances are eliminated by the
four-terminal configuration and the high input impedance of the transformer
ensures that little current flows in the potential circuit.

The stability of the necessary reference resistor, with temperature and
time, is still important, and can be checked by measuring a standard resistor.
Shielded cables are often used as thermometer leads to reduce the effects of
lead inductance and capacitance and the shields are often driven guards to
minimise the effect of leakage to ground. Resistance-thermometer sensors are
almost pure resistors with barely-detectable reactive components, at least at
frequencies below 400 Hz. However, insulator-leakage effects start to become
important at temperatures above 500°C and can introduce large reactive
components into the effective impedance of the PRT that may saturate the
bridge quadrature balance. Much work is being done to identify and counteract

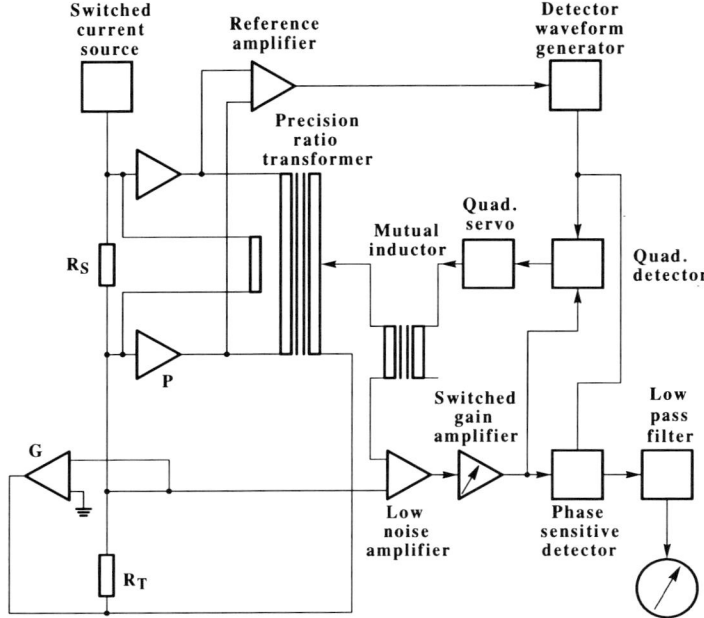

Figure 3.11: Automatic Systems Laboratories (ASL) AC bridge.

the problem. There are many instruments employing AC ratio transformers on the market now with varying applications, and prices and advice should be sought before purchasing such a device.

3.5.2   Active bridges

The use of operational amplifiers allows the design of bridge circuits that have extremely high impedances in crucial parts of the circuit, when the bridge is balanced. This, combined with the use of a ratio-transformer potential divider, has resulted in the development of the NML-designed resistance-thermometry bridge—see Figure 3.12. A controlled current source (CS) supplies current to the resistance thermometer that is proportional to the output voltage, V, of a sine-wave oscillator, operating at 80 Hz:

$$I = \frac{V}{R_s},$$

where R_s is the transresistance of the CS. The transresistance of the current source is determined by stable reference resistors incorporated within the CS.

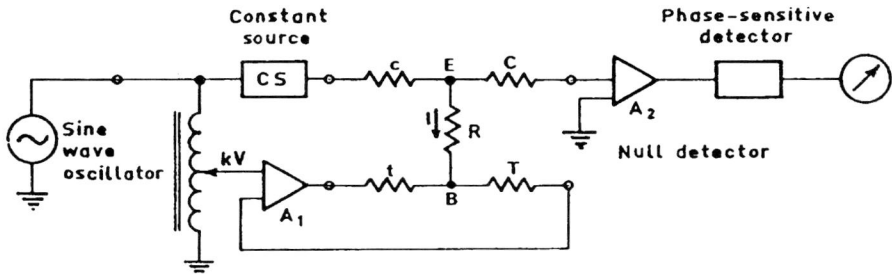

Figure 3.12: NML precision resistance thermometry (active) bridge. The circuit uses operational amplifiers to transfer precisely-set voltages to either end, B and E, of the PRT's sensor, R. The resistances of the thermometer leads are indicated (c, C, t and T) and the output voltage of the oscillator is V.

The oscillator also excites the ratio transformer, which is tapped at some ratio k, when the bridge is balanced. The tapped voltage, kV, is fed to a unity-gain operational amplifier, which in turn will drive the voltage at 'B' to the same kV. Under balanced conditions the voltage at 'E' is zero and the current I flowing through the thermometer causes a voltage drop of kV across the thermometer sensor (of resistance R).

$$\text{Hence} \qquad I = \frac{V}{R_s} = \frac{kV}{R},$$

$$\text{and} \qquad R = k\,R_s.$$

The commercial version of this bridge uses a ratio transformer for the upper three decades and a Kelvin-Varley resistance divider for the lower two. It is designed to have errors no greater than $0.01\,\Omega$ and departures from linearity in most bridges is less than $0.004\,\Omega$. Errors of this magnitude will appear mostly when the first step of a decade is switched into the circuit, (e.g., the step from 99.99 to $100.00\,\Omega$ can be different from the nominal $0.01\,\Omega$ by as much as $0.004\,\Omega$). The ohm value of the bridge depends on the resistors that determine the transresistance, R_s, and are normally adjusted to give an uncertainty of better than $0.01\,\Omega$.

The out-of-balance voltage appearing at the recorder terminals is linear until it exceeds the equivalent of about $0.3\,\Omega$ or $0.8\,^{\circ}\mathrm{C}$ on a $100\,\Omega$ R_0 thermometer. The linearity calibration of this type of bridge presents some problems, as the divider is of the Kelvin-Varley type and, in theory, the correction to any decade step is dependent on the overall setting. A procedure that allows any combination of dial settings to be compared with the $100.00\,\Omega$ step, for instance, using an external ratio transformer, has been described by

MacGregor (section 3.6.4). This method quickly allows the identification of any such error.

3.6   Calibration of instruments

The calibration of instruments in resistance thermometry normally requires only a check on their linearity because the interpolation equations for thermometers are usually expressed in terms of resistance ratios.

3.6.1   The calibration of a Mueller bridge

This section outlines a calibration procedure for resistance bridges whose adjustable decade steps are arranged in a series-addition configuration. The procedure involves two stages.

Firstly, internal values are intercompared relative to the 'bridge ohm' or the equivalent (e.g., the $10\,\Omega$) and, secondly, the bridge ohm is measured in terms of the absolute ohm. Because the resistors in any decade are of the same construction and usually from the same batch, their drift with time is usually very consistent. Consequently, if the whole of a decade ($10 \times 10\,\Omega$, say) is used to measure a standard resistor ($100\,\Omega$) and the drift is within tolerance limits, it is usually reasonable to assume a drift of $1/10$ of the total for each member of the decade. Thus, a complete 'build-up' needs to be done only once or twice per year, and even less frequently once a history of drift has been obtained. The stability of the ohm value of the bridge can be readily checked by measurement of a standard resistor. However, as far as resistance thermometry is concerned, the stability of the bridge ohm is only important for the monitoring of R_0, as temperature measurement depends only bridge linearity.

The basis of the linearity calibration is as follows.

Starting with the $10\,\Omega$ decade, say, a stable $10\,\Omega$ reference resistor (usually a decade box or build-up resistor) is connected to the bridge and the bridge is balanced using the lower decades.

Let A_1 be the correction to the first $10\,\Omega$ step, A_2 that of the second and so on to the tenth, A_{10}. Similarly, let the corrections to the $1\,\Omega$ steps be B_1 to B_{10}. The tenth step is often included in high-quality instruments as a convenience during readings and in calibration. The bridge reading on measuring the $10\,\Omega$ reference resistor will be:

$$\text{reading } 1 = 10\ \Omega - A_1 + \delta R_1 + Z\,,$$

where Z is the bridge zero and δR_1 is the correction to the nominal value of the external $10\,\Omega$ resistor. The value of δR_1 does not matter (see below) provided it is constant in the short term. The same applies to Z, which is the sum of the corrections to zero settings on the decade.

$$Z = A_0 + B_0 + C_0 + \ \ .... \ . \tag{3.4}$$

Next, the bridge is balanced using the same external resistor, but with the 10 steps of the $1\ \Omega$ dial instead of the $(1 \times 10\ \Omega)$. Then:

$$\text{reading } 2 = (10 \times 1)\ \Omega - B_{10} + \delta R_1 + Z$$

and the difference, δA_1, between the above two readings will be

$$\delta A_1 = -A_1 + B_{10}, \tag{3.5}$$

i.e., the difference between the correction to the $10\ \Omega$ and that of the $10 \times 1\ \Omega$ steps.

The external resistor is then increased to $20\ \Omega$ and the bridge is balanced using the $(2 \times 10\ \Omega)$ step and then using $(1 \times 10\ \Omega) + (10 \times 1\ \Omega)$, to give:

$$\begin{aligned}
\text{reading } 3 &= 20 - A_2 + \delta R_2 + Z \\
\text{reading } 4 &= 20 - A_1 - B_{10} + \delta R_2 + Z \\
\text{and thus} \qquad \delta A_2 &= -A_2 + A_1 + B_{10}. \tag{3.6}
\end{aligned}$$

Adding equations (3.5) and (3.6):

$$\delta A_1 + \delta A_2 = -A_2 + 2 \,.\, B_{10}, \tag{3.7}$$

$$\text{or, in general,} \qquad -A_n + n \,.\, B_{10} = \sum_{1}^{n} \delta A_n$$

$$\text{and} \qquad A_n = n \,.\, B_{10} - \sum_{1}^{n} \delta A_n. \tag{3.8}$$

Either the correction A_1 is known from an ohm measurement or it is assumed to be zero, in which case other corrections will be relative to it. Thus, the other 'A' can be calculated, as well as B_{10}. The next dial down will yield the equation:

$$B_n = n \,.\, C_{10} - \sum_{1}^{n} \delta B_n. \tag{3.9}$$

The value of B_{10} is known from equation (3.8), and so the rest of the 'B', as well as C_{10}, can be determined. The other dials can be determined in a similar way. If the bridge does not have the tenth step in the decade the same

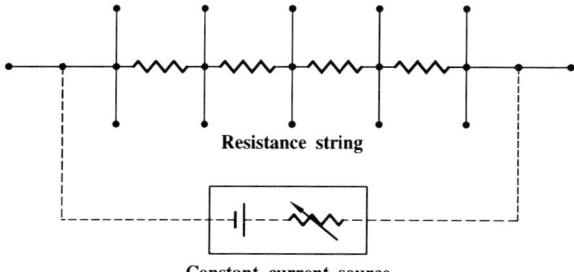

Figure 3.13: A resistance build-up group for checking the linearity of DMM's and instruments based on the Kelvin-Varley divider.

procedure is used, but the sum of all the 'nine' steps of the dials below that being measured is used. A potentiometer is calibrated by similar means.

The build-up calibration is tedious and time consuming. However, as pointed out, such calibrations are essential at least until the drift rate is small enough to warrant only occasional checks. Many resistor-based instruments will be subject to shock and therefore should be calibrated *in situ*. As far as ratio-transformer instruments are concerned it is only necessary to calibrate the internal reference resistor periodically apart from initial checks on the linearity of the divider itself.

3.6.2   DMM calibration

The manufacturers' specifications for DMM's usually make fairly generous allowances for their absolute accuracy and long-term drift. PRT measurement, as emphasised above, requires only knowledge of the linearity of the instrument and, usually, only over a comparatively small range (e.g., 100 to 300 Ω for a digital ohmmeter). NML offers a comprehensive calibration service for, at least, the voltmeter part of a DMM, but at a considerable charge. It has been the experience of the NML temperature group that most good-quality digital instruments are linear to more than a sufficient degree for the requirements of most industrial platinum-resistance thermometry. Moreover, the linearity of good-quality DVM's appears to be within the resolution on any one range.

Linearity checking

The build-up procedures for Mueller bridges are not appropriate for DMM's. Nor are they suitable for instruments that use the Kelvin-Varley divider system, as the error of any particular step depends on the settings of the higher-order dials. A full calibration would thus involve checking every

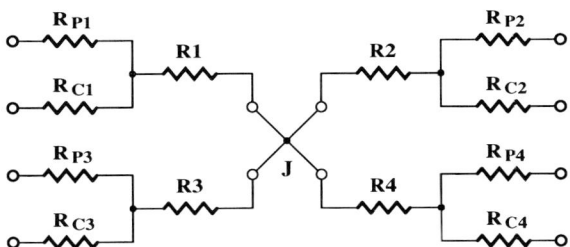

Figure 3.14: Network of 4 terminal resistors generating 35 calculable resistance values to check the linearity of AC-resistance bridges.

combination of dials and is therefore impractical. However, in practice, the biggest errors arise when a new decade is introduced into the measurement and these steps are the ones worth checking. One way of doing this is by connecting a number of resistors of suitable values in series, with a measurement branch at each junction. Figure 3.13 illustrates such an arrangement, which may be used for a DMM as well.

Each resistor is measured in turn, then in groups. A typical group might have the values $100\,\Omega$, $50\,\Omega$, $50\,\Omega$, $10\,\Omega$ and another $100\,\Omega$. By moving from either end, 100, 110, 150, 160, 200, 210 and $310\,\Omega$ are covered. Also, if one of the $100\,\Omega$ resistors is slightly below nominal, the important step 99 to $100\,\Omega$ would be checked.

3.6.3 Linearity checking of AC ratio-transformer resistance bridges

Linearity calibrations of AC ratio-transformer bridges can be carried out using the above technique. However, the procedure has been considerably refined in a commercially-available resistance-bridge calibrator using four measurable resistors that can be combined in 35 series and parallel combinations, each of which has a value calculable from the values of those resistors. The arrangement is set out in Figure 3.14 and described in reference [31].

3.6.4 Calibration of the Leeds and Northrup temperature-measuring bridge

The calibration of a five-dial bridge is often accomplished by presenting a fully calibrated six-dial decade standard resister to its terminals. The techniques are well known and laborious. The inherent accuracy and stability of the multi-dial reference ratio transformer[1] provides a much simpler solution to the problem whenever alternating currents within the audio-frequency range are involved. The following is a method used at NML for the calibration of the Leeds and Northrup 8078 Precision Temperature Bridge.

This bridge has two modes of operation: the OHMS mode, in which resistance is measured relative to a pre-set internal standard, and ICE POINT SET, where resistance is measured relative to an internal standard, which may be adjusted over a small range from the front panel.

A complete calibration consists of two parts:

1. examining the values of the bridge dial settings relative to a nominal setting (usually 100.00) and

2. determining the true value of the nominal setting in the OHMS mode.

The second part requires a suitably-calibrated standard resistor, in terms of which the bridge may be adjusted or a correction established. This will not be considered further.

For the first part, the method described uses a reference ratio transformer as a means of transforming the value of a single fixed resistor. The equivalent four-terminal resistance is measured on the bridge in the ICE POINT SET mode. The arrangement of this circuit and its connection to the bridge terminals are shown in Figure 3.15.

Current I from the constant-current source of the bridge enters the ratio transformer at its tap k, and is divided so that the fraction kI flows through the resistor R and the remainder $(1 - k)I$ passes through the low end of the transformer to virtual ground. At balance the bridge measures the voltage between the potential terminals of the resistor, kIR, relative to the voltage across the internal standard resistor, which is carrying the same current I. The purpose of the virtual-ground current-buffer amplifier is to return the current $(1 - k)I$ to the end of the resistor remote from the transformer, so that the bridge sees the same current I at both current terminals. Thus, the effective resistance seen by the bridge is kR.

Two possible sources of error need to be considered. The first is the impedance of the ratio transformer as presented to the constant-current source. This impedance can be as much as $5\,\Omega$, and therefore the bridge must be used in accordance with the instruction manual for keeping errors due to lead resistance within the specified limits. In particular, when used in the ICE POINT SET mode, the extent of the adjustment must not exceed $0.1\,\Omega$ with respect to the nominal $100.00\,\Omega$.

---

[1]Commercial five-dial ratio transformers have a ratio uncertainty better than $10\,\mathrm{ppm}$ at $80\,\mathrm{Hz}$.

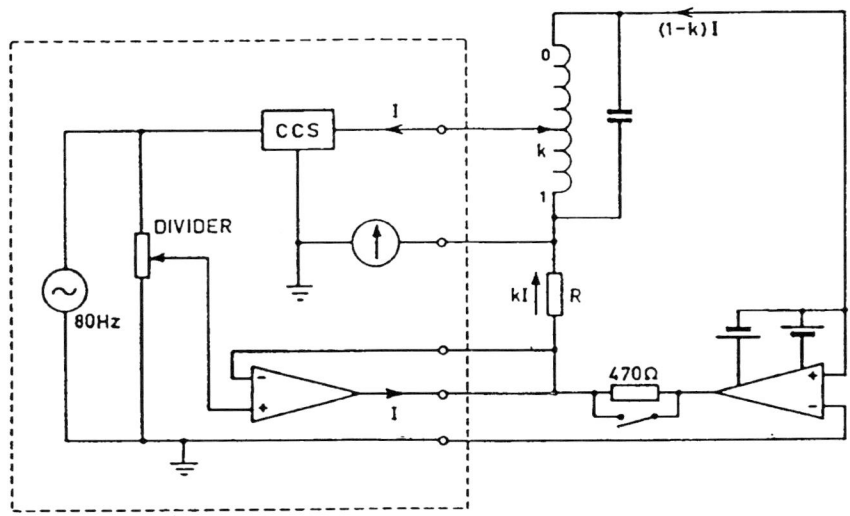

Figure 3.15: Schematic of the L&N 8078 bridge and its calibration circuit (see text).

The second source of error is that due to the finite magnetising impedance of the ratio transformer and the finite voltage gain of the virtual-earth current-buffer amplifier. The small input voltage to the buffer will cause a small magnetising current to flow in the transformer contributing to the current in the resistor R. If at 80 Hz the voltage gain of the amplifier is A and the impedance of the transformer is Z, then the error current is kIR/AZ, giving a fractional-resistance error of $1/AZ$. Since the gain of the amplifier at 80 Hz includes a phase shift of nearly $90°$, most of the in-phase component of the error can be removed by tuning the ratio transformer. The transformer is tuned when the $470\,\Omega$ resistor at the output of the amplifier can be switched in or out of the circuit without affecting the bridge balance. This adjustment is not critical and only needs to be set up once. For the type of ratio transformers used, the tuning capacitance at 80 Hz is about $0.1\,\mu\text{F}$. The complete circuit and its connections to the bridge are shown in Figure 3.16.

Operation

A value of $R = 100\,\Omega$ is used to calibrate the 10, 1, 0.1 and 0.01 Ω dials. This ensures a simple 1:1 correspondence between the bridge and the reference transformer settings.

It is preferable to read the offset of the bridge from the recorder output rather than adjust to a null, which would require extra dials on the reference transformer. A chart recorder with a range of $\pm 50\,\text{mV}$ is convenient.

The external calibrating circuit is connected to the bridge with the reference transformer set at 1.000 00, and the bridge dials set at 100.00 in the OHMS mode. The external resistor R is adjusted to balance the bridge to 0.01 ohm. The bridge is then switched to the ICE POINT SET mode. Now, with the bridge dials set at 0.00 and the reference transformer to 0.000 00, the zero offset is observed. The settings are restored to 100.00 on OHMS and 1.000 00 and the ICE POINT SET control is adjusted to give the same indication as for the zero. Thus, the increment from 0 to $100\,\Omega$ on the bridge dials corresponds exactly to the increment from 0.000 00 to 1.000 00 on the reference transformer. The chart recorder sensitivity may be calibrated by offsetting either the transformer or the bridge dials. The zero should be checked periodically, since it drifts slightly with time. Similarly the $100\,\Omega$ settings should be checked periodically.

The bridge is calibrated against the reference ratio transformer by setting both to corresponding readings and observing the error on the recorder output. Each dial in turn should be checked at all settings with the other dials set to zero, and some selected combinations of dial settings should also be checked.

The 100.00 to 200.00 ratio is checked by setting the bridge to the OHMS mode and its dials to 100.00 as before. The ratio transformer is set at 0.300 00 and the resistor R adjusted to balance the bridge at to $0.01\,\Omega$. R will be nominally $333.33\,\Omega$. With the bridge switched to the ICE POINT SET mode, the previous procedure is repeated. The 200.00 setting is checked by switching the ratio transformer to 0.600 00. Any intermediate point (in fact any point) can be examined by setting the ratio transformer to three times the bridge reading.

Calibrations of a number of bridges have shown errors up to a maximum of $0.003\,\Omega$ in the relative-resistance values. Errors at this level should therefore be considered normal for this type of bridge.

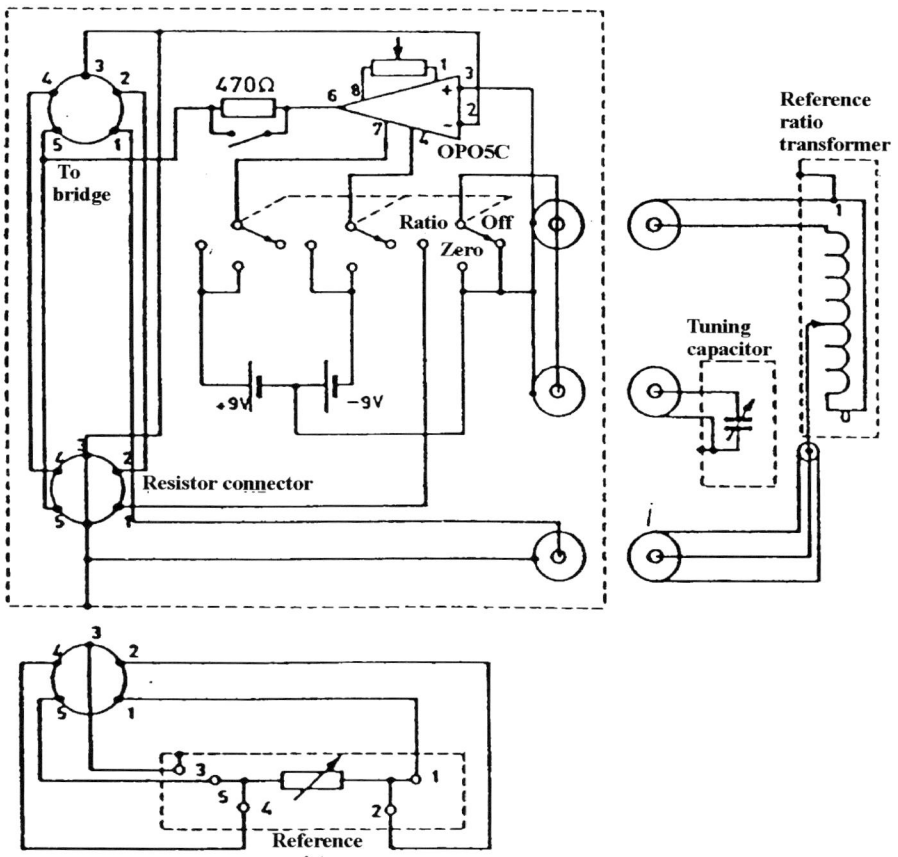

Figure 3.16: Test circuit, reference transformer and reference resistor for adjusting the L&N 8078 bridge.

Chapter 4

Digital Thermometers

John Connolly

4.1   Introduction

So far, the discussion of temperature measurement has dealt with the sensor and the measuring system as separate entities. However, there are many instruments available that really do deserve the designation *thermometer*, as they give a reading directly in temperature. This is in contrast to a platinum-resistance 'thermometer', for example, that will not provide the user with a measure of the temperature without the interaction with a resistance-measuring instrument and a method of conversion to temperature. Prior to the advent of digital processors and analogue-to-digital converters most thermometers employed an analogue indication of the measured parameter, calibrated as a temperature scale. Examples are moving-coil ammeters and, of course, liquid-in-glass thermometers, where the temperature-dependent volumetric expansion of a liquid is viewed in a capillary tube.

In a **digital thermometer**, an electrical output corresponding to the temperature-dependent property of a sensor is digitised and then converted to a temperature display, using a mathematical processor—the entire process from sensor to display being in one integrated unit. Digital thermometers come in a variety of types, varying in resolution, sensor type and price, but, in this chapter, attention will be mostly limited to those with a resolution of 0.01°C, or better, and using platinum-resistance thermometers (RTD's) as sensors. However, a brief reference will be made to other types (section 4.3).

4.1.1   Analog conversion

Older style digital resistance thermometers used the potentiometric resistance-measuring technique described in section 3.1. The temperature is obtained in a three-step process.

- The resistance of the RTD is obtained from the ohmmeter.

- Electronic scaling and linearisation convert the reading to temperature.

- The reading is displayed on an analogue meter or, in more modern instruments, on a DVM-based digital panel meter.

The conversion to temperature might be just scaling, perhaps with a zero (0 °C) adjustment, but is then limited by the inherent non-linearity of RTD's. The non-linearity might be compensated to some degree, and such instruments are usually designed to conform with a standard specification code, such as IEC–751 (section 2.2). The precision of such instruments is limited to the tolerance set out in Figure 2.1. The first development of a direct digital-display of temperature used a digital voltmeter-based panel meter.

More sophisticated (and higher-priced) instruments have linearisation in-built, with electronically-derived adjustments for slope and, perhaps, curvature. The calibration of these instruments involves a set-up procedure carried out at one, two or three known temperatures, depending on the instrument, followed by a normal intercomparison calibration against a reference thermometer. The calibration should be carried out at sufficient temperatures spanning the range of use to allow meaningful interpolation. However, it is unlikely that these instruments would be capable of measurements to better than 10 or 20 mK, and certainly no better than twice the least count. These instruments are now superseded by those using digital processing for the primary resistance measurement and for the conversion to temperature.

4.2   Digital resistance thermometers

Many modern RTD-based digital thermometers make use of high-resolution analogue-to-digital (A/D) converters incorporated into a potentiometric measurement of resistance, as described in section 3.4. The temperature is calculated from the resistance measurement using an in-built microprocessor and some form of interpolation equation, whose coefficients, specific to the particular RTD, are stored in an internal memory. Some can also display the RTD's resistance relative to that of an internal reference resistor. All high-precision instruments incorporate a procedure for entering calibration coefficients into memory, either by a front panel keying-in procedure or from a computer via a communications link.

The interpolation equations employed vary from the reference function and deviation equation of ITS-90 to polynomial relationships of resistance versus temperature, with the degree chosen to approximate ITS-90 with the required precision.

The calibration of these instruments is a two-step procedure. The coefficients of the RTD's interpolation formula are determined using a normal sensor-calibration procedure and the *thermometer* is then calibrated at sufficient temperatures to achieve the necessary interpolation precision.

The development of these instruments continues unabated. The precision and resolution of A/D converters is such that resolution in temperature display can be as fine as 10μ°C with an uncertainty better than 0.001°C, at least for temperatures below about 250°C. There are RTD-based thermometers for which all the measurement, processing and communication circuitry is packaged into the sensor's head—all that is needed to display temperature is a serial-communication cable connected to a personal computer and a little software.

4.2.1 AC ratio-transformer thermometers

The ratio-transformer resistance bridges discussed in section 3.5 are now mostly self-balancing and incorporate digital communication ports that can interface to a computer. Thus, they can constitute a digital thermometer if software is written to convert measured resistance to temperature.

There are instruments that can make use of the inherent stability of the ratio-transformer technique, with internal microprocessors, to directly display temperature with the highest-practical precision. As with the instruments described in section 4.2, a range of interpolation equations are employed and it is up to the user to choose the most appropriate one for the application. Calibration can involve the determination of coefficients for the interpolation equation, although, often, those supplied by the manufacturer are close enough and all that is required is a correction table, giving corrections with respect to the built-in equation. This is obtained from intercomparison with a reference thermometer, a process that also checks instrument linearity and the accuracy of the conversion software.

4.2.2 Range change and hysteresis

In some digital thermometers the range is extended automatically by a step change in the amplification of the analogue input to the A/D converter when the measured value has reached the top of the range. Such a change shows up by a reduction in the least count of the display. This can introduce two effects that may cause errors or, at least, confusion. Firstly, there is often a step change in the correction as a result of an inexact amplification step, even when a corresponding reduction in resolution is taken into account. Secondly, the reading at which the change occurs may differ depending on whether the temperature is rising or falling. In some thermometers the hysteresis range is as much as 20°C so that when such instruments are being calibrated the overlapping region must be identified and the corresponding errors determined.

4.2.3   Error sources

All sources of temperature-measurement error referred to in section 2.11 will have to be taken into account when assessing the uncertainty in use of a digital thermometer. However, there are a number of errors that may be of special concern and are worth repeating.

Sensor-current heating

The increase in temperature of an RTD due to heat dissipation from the measuring current was discussed in section 2.11. However, digital-resistance thermometers generally have a fixed current, so there is no procedure for determining the correction. To some extent, the self-heating effect is included in the calibration, although it is worthwhile checking the instrument's specifications to be assured that no significant error is being introduced. As a rule of thumb, a current of $1\,\text{mA}$ in a $100\,\Omega$ RTD immersed in water will cause a rise in the sensor temperature of a little less than $0.01\,°\text{C}$.

Parasitic-emf sources

Thermal- and chemical-emf sources in RTD's will cause a bias in the A/D output. Most high-quality instruments will have a strategy to remove such emf's, mostly by reversing the direction of the measuring current, and the specifications for the instrument should be checked.

Of course, spurious-DC sources have no significant affect on the operation of digital thermometers based on AC bridges. However, there are errors sources peculiar to AC systems. These include the adequacy of quadrature balancing of the bridge, shunting effects in lead insulation with poor AC properties and AC/DC resistance differences.

Reading averaging

Some digital thermometers have adjustable reading rates and sample sizes for averaging. Averaging is useful for reducing the effect of random noise with a frequency at about or greater than the reading rate. However, isolated unidirectional noise pulses will cause a bias in the reading that persists over the number of readings in the average. If this type of noise is present it is better to use a single-reading mode and deal with the noise in another way.

Ambient temperature

Many digital thermometers are small and portable, making it easy to take them into measurement locations having differing ambient temperatures, often quite different from that existing during calibration. Sometimes the calibration data contain information on the effects of changes in ambient temperature, but, more usually, reliance has to be placed on the manufacturer's specifications.

Software errors

The software used in the measurement and control of digital thermometers can introduce errors. An obvious example is the loss of pre-set data, either by inadvertent entry into an edit mode or by some corruption of it. This presupposes that the software does, in fact, have adequate resolution for the required significance in the arithmetic. Checking of the stored data should occur at frequent intervals and instruments that do not allow recall of stored data for checking purposes should be regarded with suspicion. The adequacy of internal processing should be evident from any instrument-calibration checks.

Another source of software-based error occurs in some of the digital resistance thermometers that have an internal auto-calibration procedure initiated at switch on. Part of the procedure is a measurement of reference-device temperature to correct for ambient temperature variations. Some do not appear to track subsequent temperature variations as the instrument warms up. In such cases it is necessary to re-initiate the auto-calibration after some 20 to 30 minutes, when the thermometer will have reach stable conditions.

4.3 Other digital thermometers

The sensor in a digital thermometers may be an RTD, as discussed above. On the other hand, it may be a thermocouple, a thermistor, a silicon integrated-circuit chip or, indeed, some unspecified device. Some devices have a rather limited range around ambient temperatures. Probably the most common digital thermometers are those using thermocouples as sensors. However, they have their own particular configurations, for example, the need to incorporate cold-junction compensation, and are discussed in Volume 3 of this handbook [32].

Thermistors (section 2.4) have a highly non-linear resistance-temperature characteristic, although interpolation using modern microprocessors minimises any associated disadvantage. The attraction for thermistors is their high sensitivity and small response times. An integrated-circuit sensor, usually either a temperature-dependent current source or based on a temperature-dependent zener diode, is also small and responds fast. In both sensor types excitation-current self-heating can be a major source of error and is often difficult to assess. To some extent it can be calibrated out, but tests in a variety of sensor environments with varying thermal resistances should be carried out.

Chapter 5

Liquid-Expansion Thermometers

Corinna Horrigan

5.1   Introduction

The earliest recorded temperature measuring devices date back to Graeco-Roman times and used fluid expansion to measure temperature changes. These early instruments were very basic. They had no scale and, as they were open to the atmosphere, they were strongly affected by air pressure as well as temperature. The renewed interest in natural science during the Renaissance encouraged the development of both instrument and scale. At one time thermometers were quite fashionable and many strange instruments were created. However, this did not impede the development of the liquid-in-glass thermometer as a practical scientific instrument. Modern instruments made for specialised areas of measurement, such as deep ocean temperatures, are every bit as complex as the fantasies of the 17th century, as illustrated by Figure 5.1. An excellent summary of the development of the liquid-in-glass thermometer is given by Middleton [33].

Despite the introduction of many other types of thermometer, liquid-in-glass thermometers are still the most familiar of temperature-measuring devices. They are cheap, portable, easy to read and require no auxiliary equipment and so are found in many homes. One result of this familiarity is that they are often used incorrectly. Many users assume that there are no difficulties involved in getting a correct measurement from a liquid-in-glass thermometer when they might be more sceptical with a less familiar instrument.

In fact, the liquid-in-glass thermometer is a complex system of interacting

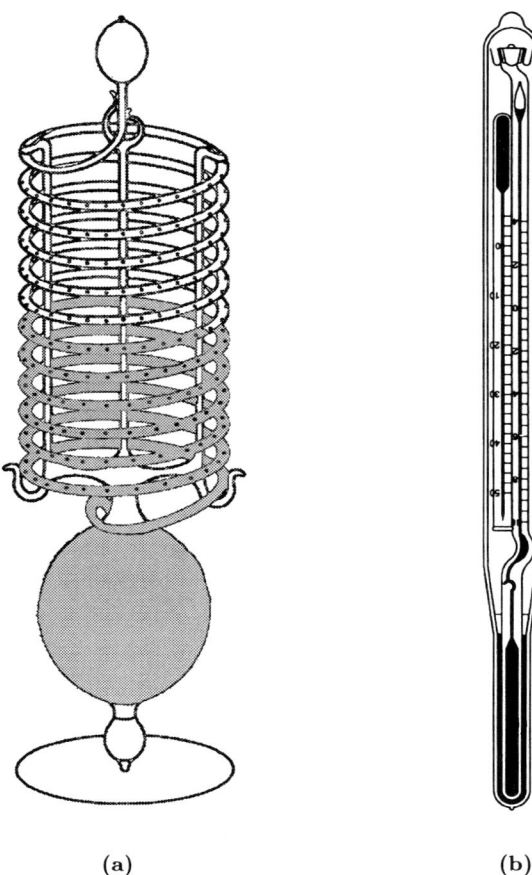

(a)   (b)

Figure 5.1: Two unusual thermometers: a Florentine thermometer (left) circa 1600 and a modern deep-sea reversing thermometer (right).

components and substantial errors can arise if it is not properly used. On the other hand, it is a mistake to assume that they are grossly inaccurate instruments. The achievable accuracy depends on the individual thermometer and the experience of the user. For many decades, these were the only temperature-measuring devices available for scientific research and a lot of effort was expended in understanding and developing the instrument so that high accuracy could be achieved when they were correctly used. As a result, thermometers are available that, if properly calibrated and used, can achieve uncertainties as low as 0.002 °C in the measurement of temperature differences.

This chapter deals with the basic principles of liquid-expansion thermometry, and Chapters 6 and 7 cover calibration methods, correct care and use of liquid-in-glass thermometers, uncertainty and special types of thermometers. Liquid-expansion thermometry covers more than just the liquid-in-glass thermometer, but as this is the most common type of expansion thermometer, other types are covered only briefly, in Chapter 7.

5.2 The liquid-in-glass thermometer

5.2.1 General description

Most materials expand when heated and contract again when cooled. Liquidin-glass thermometers make use of this property to measure temperature. Unfortunately, the process is made complex because more than one material is involved and because few materials expand linearly over their whole range.

A typical liquid-in-glass thermometer is shown in Figure 5.2. It consists of two main sections: the bulb and the stem. The bulb is a thin-walled glass cylinder that contains most of the thermometric liquid. Some of the liquid, however, extends along a fine capillary, which is attached to the open end of the bulb. The section that contains the capillary is called the stem. As the thermometer is heated the liquid expands further along the capillary, and it contracts towards the bulb when cooled.

The size and shape of the bulb affects the sensitivity and performance of the thermometer. The more liquid that is contained in the bulb, the greater will be its expansion along the capillary when heated, and the more surface area that is exposed to the temperature to be measured, the smaller the response time. Thinner walls also reduce the response time, but they increase the fragility of the instrument and decrease the stiffness of the bulb, which may result in increased stiction (section 5.3.5). The type of glass used in the bulb is also an important factor in thermometer performance and will be discussed in more detail in section 5.2.4. All of these factors, and others, such as bore diameter and the expected use of the thermometer, must all be considered in thermometer design.

The stem may be of either the 'solid-stem' or the 'enclosed-scale' variety. A solid-stem thermometer has a capillary tube with a very fine bore, but quite thick walls (up to 6 mm). This gives the stem extra strength and also allows the scale to be permanently etched directly onto the stem. The main drawback is the increased possibility of parallax errors on reading the thermometer. On the other hand, the scale of an enclosed-scale thermometer is printed onto a separate backing plate, which is clamped onto a thin-walled capillary tube, and both are surrounded by an outer glass sheath. One advantage of this design is that parallax problems are reduced, as the scale and meniscus of the liquid are very close together. But the disadvantages are increased fragility

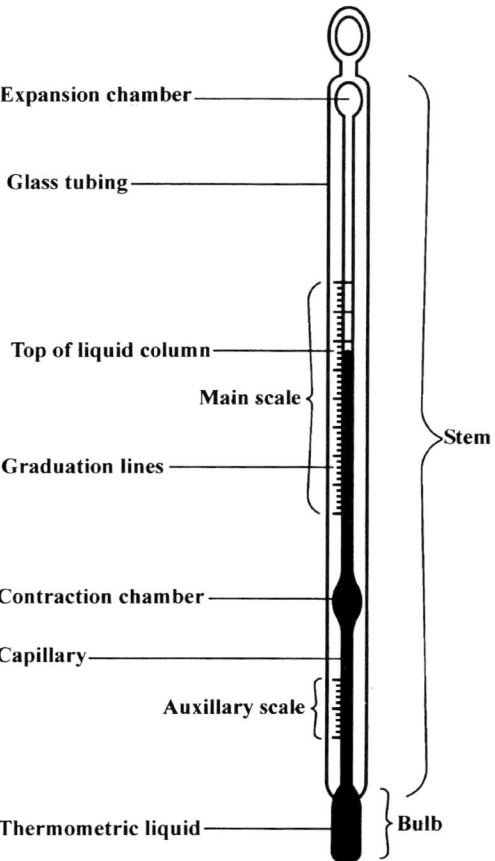

Figure 5.2: A typical solid-stem liquid-in-glass thermometer.

and the possibility that the scale may become detached from the capillary and result in faulty readings. Good thermometers of this type usually have a datum line etched on the capillary so that such displacements of the scale are apparent.

The bulb and capillary form a sealed container and the space above the liquid is either vacuous or filled with a dry, inert gas. The expansion chamber at the top of the capillary allows for gas compression and gives a slight safety margin for overheating. In volume, this chamber should be equivalent to at least 20 mm of straight capillary. To avoid trapping the liquid, the chamber is usually an inverted-pear shape. In thermometers made from a type of glass that is hard to work, the capillary may simply be extended for a sufficient length.

The contraction chamber is a widening of the capillary in the main body of the thermometer, usually below the main scale. It can serve two purposes. Firstly, it can shorten the overall length of the thermometer, so making it more manageable—an expansion, which might take up several centimetres of normal capillary, will be equivalent to only a few millimetres in height of the contraction chamber. Secondly, the contraction chamber prevents the liquid from retreating completely into the bulb, a situation that allows gas bubbles to be trapped in the bulb and these can be difficult to remove.

Some thermometers have two scales, the main scale defining its working temperature range and an auxiliary scale, separated from the main scale by a contraction chamber. The auxiliary scale allows a particular fixed point (e.g., an ice point or a steam point) to be monitored, even though it is not part of the main range of the instrument. Many different temperature scales have been developed over the years, but the Celsius scale is the most used today and the °C is the SI unit for temperature (see Chapter 9).

The different materials that make up the thermometer—the glasses, the liquid filling and the gas filling—should not chemically interact, but they do form an interacting physical system. It is impossible to heat the liquid without heating the glass and the gas and these also respond to temperature changes, not always in an ideal fashion. The factors affecting the performance of the thermometer are discussed in the following sections.

5.2.2   Construction

The glass section of a thermometer can be made by blowing a bulb in suitable capillary tubing, but it is now more usually done by joining special bulb tubing to the capillary tubing. This results in the bulb walls being more uniform, than they would be if blown from the capillary, and allows the bulb and the capillary to be made from different types of glass. Then, a more stable glass may be used for the bulb.

After the bulb is attached and sealed at the base, the resulting blank is annealed at a temperature of about 50°C below the softening temperature of the glass. It is kept at this temperature for some time and then cooled slowly. This is an important step in the manufacturing process, as it relieves most of the strain in the glass. If it is not done correctly, then the stability of the finished thermometer will be impaired and, in extreme cases, excess strain may cause the thermometer to break as the thermal stress increases. A thermometer cannot be properly annealed after being filled with liquid and sealed as the temperatures involved are beyond its working range.

The blank is evacuated, filled with the appropriate amount of thermometric liquid and a gas filling is introduced if required. The thermometric liquid is then drawn into the bulb by cooling and the top of the thermometer is sealed. The thermometer is then cycled slowly from ambient to its upper temperature of use a number of times.

As a first step in putting the scale on the thermometer, it is 'pointed' at a number of different temperatures in its range (e.g., 0, 50, 100°C) by marking the mercury level when the thermometer is at these temperatures. The distance between pointing marks is then subdivided linearly using a dividing machine. Because the bore of the capillary is not necessarily uniform, frequent pointing is required to keep the thermometer reading close to the true temperature over its whole range. Most good thermometers have pointing marks at about every hundred divisions, though since the uniformity of capillary tubing has improved over the years, some thermometers have only two pointing marks.

The manufacture of some types of thermometer, for example clinical thermometers, can be automated, but most precision thermometers are individually made and require individual calibration.

5.2.3 Thermometric liquids

An ideal thermometric liquid would have the following physical and chemical properties:

- be liquid over at least the temperature range of the thermometer,

- have a linear coefficient of expansion over its liquid range,

- be opaque or coloured, for ease of reading,

- not wet glass, so it moves smoothly and does not coat the glass,

- be chemically inert with respect to the other materials in the system

- be chemically stable, so it does not deteriorate over time,

- be non-poisonous, for ease of manufacture and safety in use and

- have a definite and convex meniscus, for ease of reading.

Of course, no liquid fits these criteria perfectly. Mercury, however, comes closer to this ideal than any other material available. It is liquid over a wide and useful range of temperatures, its coefficient of expansion is fairly linear, it does not wet glass, it has a clearly defined, convex meniscus and is opaque. Although it oxidises in the presence of water or oxygen, it is chemically stable under an inert atmosphere. As an added bonus, it is relatively easy to purify to thermometric requirements and is not exorbitantly expensive.

Table 5.1: Common thermometric liquids and their relevant properties.

Liquid	M.P. (°C)	B.P. (°C)	Range (°C)	Problems
Mercury	−38.9	356.7	−38 to 600	Distills at high temp.; a cumulative poison.
Hg 8.7% Tl	−55	?	−50 to 20	Similar to mercury, but more poisonous; prone to oxidation.
Xylene	−47	139	−80 to 50	Non-linear expansion, wets glass, flammable and poisonous.
Toluol (toluene)	−95	110.6	−80 to 50	Similar to xylene.
Alcohol	−114.9	78.3	−100 to 20	Similar to xylene.
Pentane	−200	50	−200 to 20	Similar to xylene and cannot be dyed when used at low temperature.

Although its upper temperature of use can be extended above the boiling point by pressurising, its use below 0°C extends only to about −38°C, a limit that may be extended to −50°C by alloying with thallium (see Table 5.1). Beyond this, other less satisfactory liquids must be used. Unfortunately, mercury and most other thermometric liquids are poisonous to some degree.

The other common thermometric liquids are organic compounds. Such liquids are often quite non-linear in their response to temperature change and this is reflected in scale markings, which are more open at higher temperatures. These materials also wet glass and this has two major consequences. Firstly, the capillary of such a thermometer cannot be as fine as that of a mercury-in-glass thermometer and, secondly, some liquid will be left on the glass, as the thermometer cools, and will drain down the walls of the capillary reducing the reading over time. The problem is exacerbated when the viscosity of the liquid increases at low temperatures.

A summary of thermometric liquids, giving their ranges and drawbacks, is given in Table 5.1.

5.2.4   Thermometric glasses

Glass is an unusual material. It is a solid with an irregular or amorphous molecular structure more like that of a liquid than the regular structure found in most inorganic solids. The structure also allows the molecules some degree of mobility within the material without changing its characteristics. For these reasons, it is often described as a supercooled liquid, though it still retains some essentially solid properties, such as elasticity and brittleness. However, its unusual structure does give it some unusual properties, which affect its thermal response and thus the performance of a liquid-in-glass thermometer.

There are many different types of glass available, but not all of them are equally suited to thermometric work. Over the years special thermometric glasses have been developed with increased stability and higher softening temperatures. Even so, glass in a thermometer is still subject to long-term drift, which causes secular change, and to a short-term variation due to heating, which results in temporary depression of the reading. These are discussed in detail in sections 5.3.1 and 5.3.2 respectively.

Changes in the glass affect both the bulb and the stem, but changes in the bulb are more significant, because the bulb contains a far greater volume of thermometric liquid than the stem. There is the equivalent of about 6000 °C of mercury in the bulb of a mercury-in-glass thermometer compared to, at most, a few hundred degrees in the stem. Most stem changes can be ignored, but changes in the bulb can significantly affect the reading. For this reason the bulb is often made from a different, more stable glass type than that of the stem. Even if they are made of the same type of glass, the bulb and stem are usually made from different tubing, and great care must be taken to ensure a smooth join that will be able to withstand the thermal stresses involved and will not have projections, which may trap gas bubbles.

A thermometric glass must have the following properties:

- softening temperature at least 70 °C above the max. temperature of use,

- small and linear coefficient of expansion,

- good long-term stability after annealing,

- little short-term variability,

- homogeneity,

- no distortion of the capillary and

- no devitrification, cloudiness, striae, impurities or similar defects, which may affect the ease or accuracy of reading.

The importance of each of these properties depends on whether the glass is to be used for the bulb or the stem. For instance, a distortion in the glass of the stem may affect the visibility of the reading, but would have less effect if it were in the bulb, though it might still cause strain-related problems. Conversely, the long-term stability of the bulb is more important than that of

Table 5.2: Common thermometric glasses and their upper temperature limit.

Glass Type	Upper Limit (°C)
Normal lead glass	400
Jena normal	430
Borosilicate	460
Supremax (silica)	595

the stem, though if a change in the stem becomes too large it will be noticed. In addition, the glass of the stem should give a clear, fine line when etched.

A list of some common thermometric glasses is given in Table 5.2. However, it should be noted that other glasses may also be suitable.

5.2.5   Gas filling

Apart from some special thermometers (e.g., Beckman thermometers), which need to be vacuous, gas filling is highly recommended in mercury-in-glass thermometers. For high-temperature thermometers it is a necessity. Gas filling in thermometers with a range above 150°C decreases the possibility of distillation of the mercury, by reducing the diffusion of mercury vapour from its surface to cooler regions of the thermometer, where it may condense. In extreme examples, improperly pressurised thermometers will show a distinct drop in reading due to mercury distillation, when left at an elevated temperature for a period of time, and to re-condensed mercury droplets in the bore and in the expansion chamber. The pressure within an ordinary mercury-in-glass thermometer will usually be less than 2 atmos. but could increase to 15 to 20 atmos. in a high-temperature type, e.g., those used above 300°C. The pressure also raises the boiling point of mercury, allowing some thermometers to be used above 400°C.

In addition to reducing distillation, the pressure exerted by the gas filling also reduces the incidence of separation of the mercury column. Although, if it still occur, the filling can make the fault more difficult to rectify.

5.3   Intrinsic problems

5.3.1   Secular change

As mentioned is section 5.2.4, some changes still occur in the bulb over time despite careful annealing during manufacture. This long-term drift is

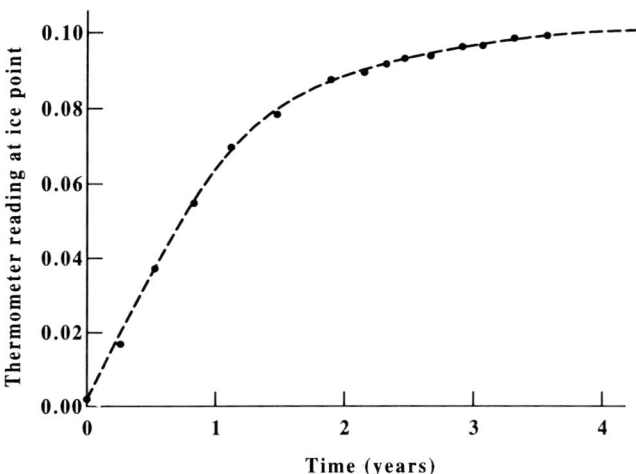

Figure 5.3: Secular change in a thermometer as a function of time, indicated by a change in the reference-point reading (measured value at the ice point).

called secular change and continues, at a decreasing rate, for the life of the thermometer.

The working of the glass during manufacture inevitably leaves some strain in the bulb—the strain is reduced by extensive annealing, but not entirely eliminated. Due to the nature of glass, the residual strain is gradually relieved, as the thermometer ages, causing the glass to shrink, the bulb volume to reduce and the level of thermometric liquid to rise. The change may be hastened by annealing at the thermometer's maximum temperature of use, but it cannot be reversed.

While secular change is continuous, it is not linear, and most of the change occurs while the thermometer is quite new—usually in the first year. The changes in a new thermometer may be very large if the thermometer is not properly annealed and, even after proper annealing, drifts in calibration equivalent to some divisions are not uncommon. The changes in an older thermometer may be smaller than the uncertainty of calibration. As the strain is relieved, the forces causing the bulb to shrink weaken and the changes asymptotically approach zero. The effect of secular change is shown in Figure 5.3.

It should be remembered that, as the change is due to an irreversible contraction of the bulb, the reading of the thermometer will increase *at every* temperature of use, by the amount of the secular change. As a consequence, the extent of secular change must be monitored over the lifetime of the instrument, by taking regular readings at a reproducible reference temperature. The most commonly used and most easily-realised temperature for this purpose is the ice point and most good thermometers will have a section of scale around 0°C, even if it is not part of the main range.

The method for correctly monitoring the reference point is detailed in Chapter 6.

5.3.2 Temporary depression

The measurement of secular change can be complicated by the existence of another effect known as 'temporary depression of zero'. After cooling from a higher temperature, most materials quickly contract to their original size. Glass is not so well behaved. An increase in temperature not only causes expansion of the glass, but the increased mobility of the components of the material may also result in structural changes. If the glass is cooled slowly, the changes have time to reverse and the glass returns almost to its original size by the time cooling is complete.

However, if cooling occurs quickly then the structural changes are virtually 'frozen-in' and, with reduced mobility, take a long time to return to their original positions. Consequently, a glass thermometer bulb will be temporarily larger than its equilibrium volume and the reading of the thermometer will be low. The effect, usually observed at the ice point, is termed temporary depression of zero, and its magnitude increases as the temperature to which the thermometer was heated is increased.

It may take days or even months for the reference-point reading to return to its original value (see Figure 5.4) and the time taken is not predictable, though in general the larger the temporary depression the longer it takes to recover.

Temporary depression has the opposite effect on the reference-temperature reading to that of secular change, but it is neither permanent nor does it affect all readings equally. For a thermometer heated to 200°C, the effect of temporary depression on the ice-point reading would be greater than its effect on a reading at 100°C. If re-heated to 200°C, then no difference will be observed as the bulb will be in its equilibrium condition for that temperature.

If we wish to measure the extent of secular change, then we must either wait the days or weeks required for the thermometer to recover or we can evolve strategies for preparing the thermometers, which ensure that the two effects can be distinguished. This is possible because the magnitude of the temporary depression after heating to a set temperature is a constant for any one thermometer, once it has become reasonably stable. To allow meaningful measures of secular change to be taken, preconditioning methods can be

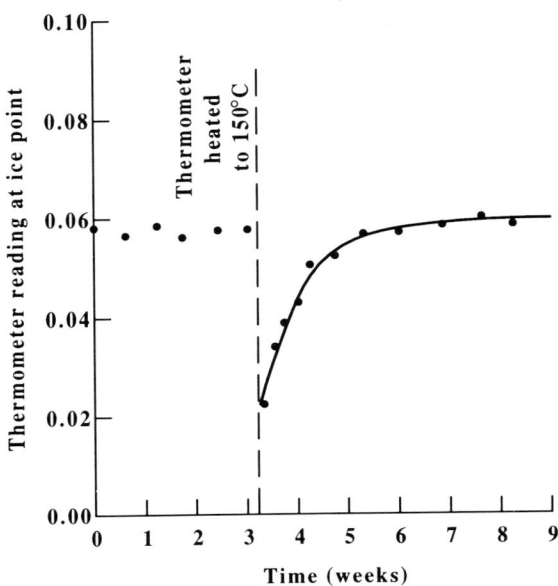

Figure 5.4: The temporary depression caused by taking a well-aged thermometer to 150°C and the recovery thereafter, as evident in measurements at the ice point.

devised to minimise temporary depression or ensure that it is a constant when the reference temperature is checked. Conditioning methods are discussed in more detail in Chapter 6.

5.3.3   Immersion effects

A liquid-in-glass thermometer is usually used to measure the temperature of a fluid. To do this it must be immersed in the fluid. The extent of the immersion depends on the type of measurement required, the depth of the fluid and the type of container holding it. There are three categories of immersion: complete, total and partial, shown in Figure 5.5.

complete immersion The thermometer is wholly immersed in the fluid with no section exposed.

total immersion The thermometer is immersed to within a few divisions of the top of the column of thermometric liquid, leaving the gas-filled section and a few millimetres of thermometric liquid exposed.

partial immersion The thermometer is immersed to a fixed distance from the bottom of the bulb. All the thermometric liquid above this point is exposed as is the gas-filled section.

A thermometer should have its immersion type clearly marked and it should be calibrated and used at that immersion as far as possible. The immersion affects the response of the thermometer, its accuracy and its ease of use. In most instances, it is not possible to use a complete-immersion thermometer because either the container or the fluid or both are not transparent. As a result most thermometers are either of the partial or total-immersion variety.

At first glance, partial immersion appears to be the easiest condition of use. The thermometer is immersed to the set distance and the temperature can then be read as required. Unfortunately, it is rarely this simple. The proportion of the liquid above the fluid, called the emergent-liquid column, is exposed to the temperature of the air surrounding it and so may be at a significantly different temperature to that of the immersed thermometer. In such a case its expansion (or contraction) will be less than if the whole column were at the same temperature. This is not a problem if the temperature of the emergent-column is the same as it was during calibration, but this is not necessarily the case. If the emergent-stem temperature is significantly different to that during calibration, then an extra correction (called the emergent-stem correction) must be added to the reading. To do this, the emergent-stem temperature must be monitored both during calibration and during use.

The magnitude of the emergent-stem correction is affected by a number of factors. The proportion of liquid in the emergent stem is one obvious factor, the change in the emergent-stem temperature between calibration and use is another and the coefficient of expansion is a third. This last is actually the differential coefficient of expansion of the liquid in glass. These factors combine to give the following equation for the emergent-stem correction Δ:

$$\Delta = K n\, \delta t$$

where K is the differential coefficient of expansion of the liquid in the glass, n is the length of exposed column, expressed as the equivalent number of degrees of scale marking, and δt is the change in the temperature of the emergent stem between calibration and use.

The significance of the emergent-stem correction depends on the accuracy of the thermometer. An emergent-stem correction of $0.01\,°C$ would be insignificant for a thermometer with an uncertainty of $1\,°C$, but highly significant if the uncertainty had been $0.005\,°C$. The methods used for monitoring emergent-stem temperature and calculating the corresponding corrections are covered in detail in Chapter 6.

Clearly, complete immersion is the ideal situation with the whole thermometer at virtually the same temperature it had during calibration, so no

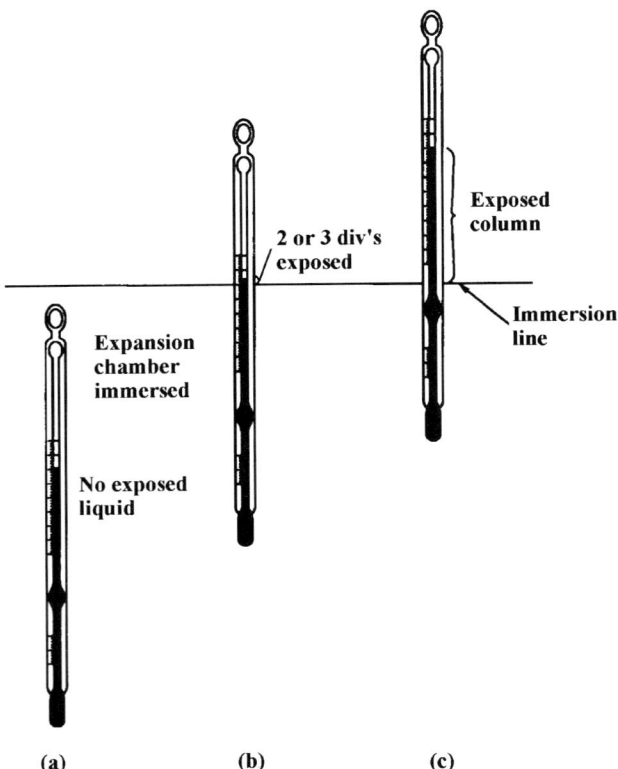

Figure 5.5: Thermometer immersion types: (a) complete immersion, (b) total immersion and (c) partial immersion.

extra corrections have to be made. However, with its use being precluded in most situations, 'total immersion' is the next best situation.

At total immersion, the bulb and nearly all of the liquid column is at the temperature of the fluid to be measured. The small amount of thermometric liquid exposed should not be different enough in temperature to cause a significant difference to the reading. This is especially so with mercury as it is a good thermal conductor. There may be a problem, however, from a change in the gas pressure in the expansion chamber, as discussed in section 5.3.4.

Total-immersion thermometers can be used at partial immersion if the appropriate corrections are calculated and applied, but they should not be used at complete immersion as the added pressure in the expansion chamber may significantly affect the reading.

5.3.4   Pressure effects

The thermometer reading can be affected by both external and internal pressure changes. The thin-walled bulb is quite elastic, so pressure changes can cause it to expand or contract.

The external pressure on the bulb is a combination of the barometric pressure and the hydrostatic pressure exerted by the liquid in which the thermometer is immersed. A pressure increase of 1 torr (133.3 Pa) will cause an increase in the reading of about 0.0001 °C. Thus, it would require a change of about 135 mm in the immersion of the thermometer or a 10 torr change in barometric pressure to cause a 0.001 °C change in the reading. For most thermometers such changes are not significant and are covered by the uncertainty of calibration.

A more sensitive thermometer may be significantly affected if it is used at a higher altitude than that at which it was calibrated. However, the change would affect all readings equally and so would be compensated by the secular-change correction. The thermometers for which external pressure is important during calibration and use are:

unprotected deep-sea thermometers that are designed to register and measure pressure and

bomb-calorimeter thermometers, designed for differential use, that are used for absolute temperature measurements at altitudes different from that at which they were calibrated.

The internal pressure exerted by the gas filling in a thermometer suppresses boiling and reduces distillation, however, it also can affect the performance of the thermometer. As the liquid expands, the gas is compressed and exerts a force on the liquid and the glass vessel. This causes minute changes in the volume of the bulb resulting in a reading slightly lower than might be expected. The effect is extremely small and usually accounted for in the pointing and the calibration of the instrument. It can become significant for very accurate thermometers (divisions of about 0.01 °C) or for those where the expansion chamber is at a very elevated temperature compared to what might be expected, e.g., the steam point in a partial-immersion thermometer.

There is also an internal pressure effect from the hydrostatic pressure due to the liquid in the thermometer. This is only really significant for mercury thermometers because of the density of mercury. Even then, it is only a problem if the thermometers are calibrated vertically, as is usual, but used in the horizontal position. This can cause a change in reading of about 0.02 to 0.03 °C depending on the length of the column at the temperature being measured. The extent of this effect can be determined, if necessary, using special calibration procedures.

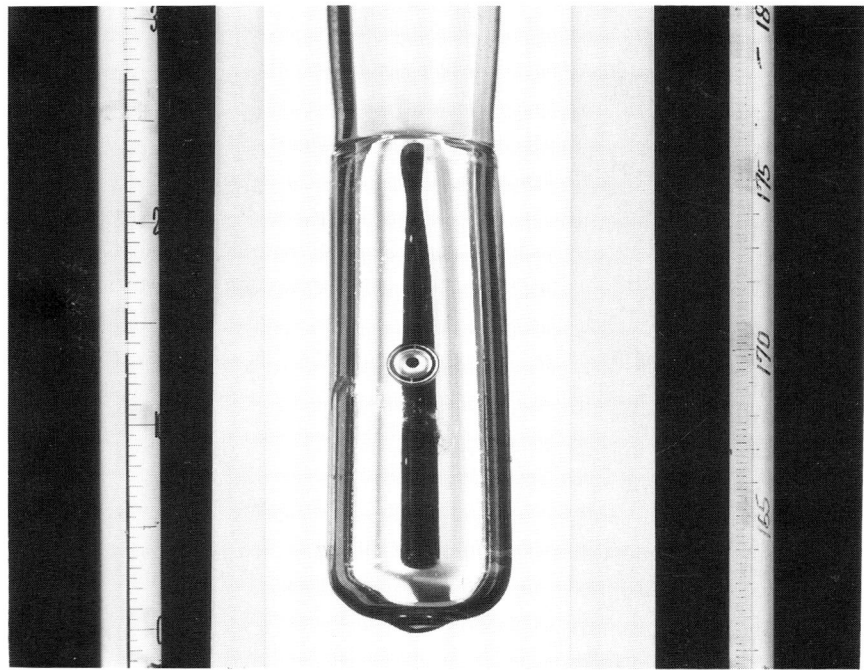

Figure 5.6: Typical examples of thermometer faults that can be rectified (from the left): broken column, bubble in bulb and worn pigment.

5.3.5   Stiction

Stiction is the tendency of the liquid to stick in the capillary. When heated, the mercury meniscus first becomes more convex and this increases the capillary pressure transmitted to the bulb. The bulb expands slightly until the restoring force becomes great enough to overcome the pressure. The mercury then rapidly moves up to a new level, the meniscus flattens again and the whole process is repeated. In the worst case, the meniscus appears to move in a series of small jumps, and is more likely to occur in thermometers with a fine bore, as this increases the capillary pressure. A thin-walled bulb also contributes to the problem, as it has less stiffness and it takes longer for the restoring force to build to a sufficient level. A fine bore and a thin-walled bulb are commonly found together in thermometers meant for high-resolution work and so it is these that are most prone to stiction.

Good thermometer design can minimise the problem, but instruments such as bomb-calorimeter thermometers must be monitored for this problem during calibration. The problem cannot be eliminated, but its effect can be minimised by gentle 'dumping' or tapping just before a reading is taken.

The magnitude of the effect depends on the individual thermometer, but has been known to be as much as one scale division.

5.3.6 Construction faults

Many thermometer faults occur through mistakes in construction or in handling. Both types will be dealt with here. Faults can be of two kinds: those that are fixable and those that are not. The rectifiable ones are usually due to handling problems that have affected the distribution of the thermometric liquid. As a result, they can be quite easily mended, and the method for doing this is described in the next chapter. These faults are quite common and include:

- broken liquid column,

- gas bubble in bulb,

- drops of liquid in expansion chamber or bore and

- worn pigment in scale markings.

Examples of these faults are shown in Figure 5.6.

Most of the other faults arise in manufacture (see examples in Figure 5.7). Thermometers with such faults cannot be mended and, in most cases, are not worth the cost of calibration—it may be possible to use them in a restricted role, but in general they should be discarded. Commonly encountered problems are:

- Uneven divisions, possibly due to a fault in the dividing process, either in the machine itself or in its placement, or it could be due to a change in the bore size. For instance, if the 100-division distance between pointing marks A and B is different to that between B and C, then the graduations between A and B will be a different size to those between B and C, and the difference will be most noticeable at B. Faults in the placement of the divider are also most noticeable at the pointing marks.

- Oxidation of mercury, caused by the entry of water or air. The oxidised mercury will form rings on the inside of the bore and will interfere with the smooth movement of the mercury and may also trap small gas bubbles. Furthermore, if the surface of a broken mercury column is oxidised, then it is not possible to rejoin the broken column.

- Thick scale lines, resulting from the etching process. If the lines are too thick, they will reduce the accuracy of the thermometer, as they can obscure the reading and tend to make the divisions appear smaller.

- Excessive strain in the glass. Strain, although it is not visible to the unaided eye, can affect the stability of the thermometer and its performance at high temperatures. A little strain may be acceptable, but excessive and highly localised strain is undesirable.

- Foreign matter in the bore, usually glass, may occasionally be seen. These can trap gas bubbles, impede the movement of the mercury, distort the meniscus or change the reading depending on where they occur and whether they can move.

- Distorted capillary.

- Missing divisions.

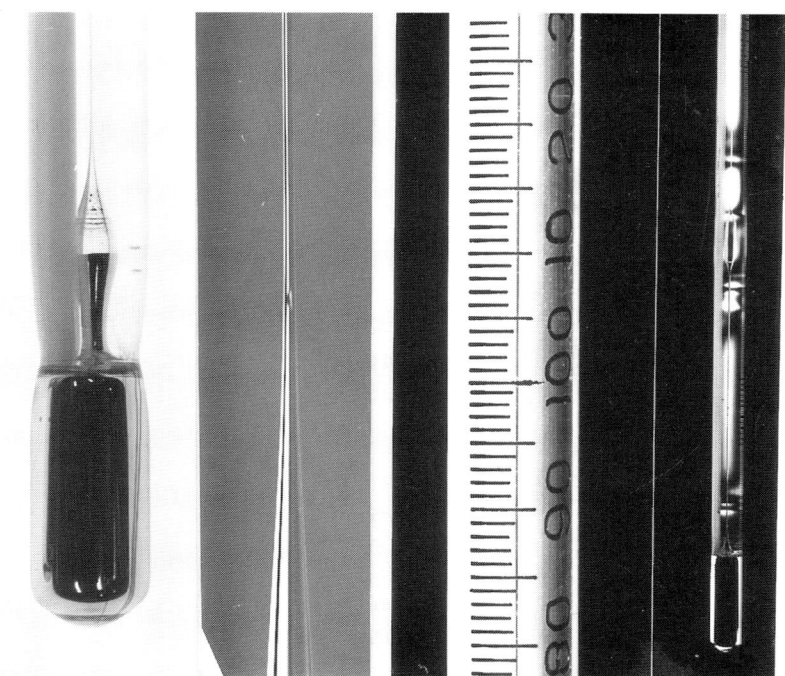

Figure 5.7: Examples of irreparable faults (from the left): oxidation, chip in bore, uneven divisions and strain.

Chapter 6

Calibration and Use of Liquid-in-Glass Thermometers

Corinna Horrigan

6.1 Introduction

The basic principles of liquid-in-glass thermometry are discussed in Chapter 5, along with some of the problems that result from the dimensional instability of glass. Before accurate temperature measurements can be made using these instruments, the accuracy of the scale placement must be checked and any changes that have occured since manufacture must be determined. This is done by calibrating the thermometer. Moreover, as these thermometers tend to change over time the change should be monitored and its causes understood if the best results are to be obtained. This chapter discusses the calibration and use of common laboratory thermometers.

The calibration of liquid-in-glass thermometers is best carried out by comparison with a calibrated standard thermometer and measurements at the ice point. At the highest level of accuracy, a standard platinum-resistance thermometer (SPRT: see Chapter 1) is used, calibrated directly at the defining fixed points of ITS–90 (Table 9.1). At lower levels, other types of thermometer may be used as standards, if they have traceable calibrations and sufficiently low uncertainties of calibration.

Calibrations at the CSIRO National Measurement Laboratory (NML) are done using an SPRT, and it is this method that is described in detail in this chapter, although a brief description of a method using liquid-in-glass

107

thermometers as standards is included. All aspects of calibration are covered, from receipt of the thermometer to the final report, including a section on the determination and interpretation of uncertainty. Also discussed is the correct use of a thermometer and understanding and using calibration reports.

6.2 Calibration

6.2.1 Examination and assessment

A thermometer submitted for calibration should first be carefully examined for construction and handling faults, of the type described in Chapter 5. A visual inspection of the capillary and of the bulb is carried out, initially using the naked eye or a low power magnifying lens and then using a microscope, if necessary. This identifies faults such as separation of the mercury column, oxidation and gas bubbles and any faults in the glass that might impair the reading of the thermometer. The scale is also examined for thick scale lines, uneven divisions and missing graduations. Sometimes a divider is used in this examination, to determine the relative lengths of portions of the scale, if the uniformity of the capillary is in doubt. Finally, the thermometer is examined using a polariscope to check the glass for signs of excess strain. Information on the construction and use of a polariscope can be found in appendix A of reference [34].

Rectifiable faults (see section 5.3) are fixed before calibration, and thermometers that can't be fixed are rejected.

While the thermometer is being examined, various relevant particulars are noted and recorded:

- required depth of immersion,
- scale reading (temperature) equivalent to required immersion,
- specification (if any),
- temperature range,
- size of graduations, e.g., 0.1 °C,
- type of gas filling,
- type of thermometric liquid,
- serial number,
- manufacturer and
- any other information that may be relevant to the calibration.

The required immersion depth, if any, is usually marked on the thermometer. If it is to be tested at a different immersion, then this is also noted.

The temperature equivalent of the immersion depth is the temperature that would be marked at the immersion line if the thermometer were graduated down to there. It is necessary to know this for partial-immersion thermometers so that the emergent-stem correction can be calculated.

A specification is a published set of constraints to which a thermometer may be manufactured. It defines various dimensions, the size of the graduations, the temperature range and acceptable corrections. A list of common specifications is given in section 6.5.3.

Also recorded is any information received from the user regarding special conditions of use or restrictions on the temperature range. This information is used to assess the temperatures at which the thermometer will be calibrated, the conditioning method to be used and the probable uncertainty—though the final uncertainty can only be determined once the calibration is completed.

Pointing marks are applied to the column during manufacture for a specific set of test temperatures, and the scale is added later using a mechanical divider. Consequently, scale irregularities are most likely to occur at pointing marks, though they are by no means unknown in other parts of the scale— between the pointing marks, the capillary may change significantly. So, in general, a thermometer should be calibrated over its entire range, with test points at and between pointing marks, or every 50 divisions, whichever gives the greatest number of calibration points. More test points would further improve knowledge of the thermometer's characteristics, but this is not really feasible and there are other sources of uncertainty that overwhelm any gains that may be made this way.

6.2.2 Reference temperatures

As explained in Chapter 5, secular change is a continuing change of bulb volume and so affects all readings. It is necessary to monitor the magnitude of secular change, after calibration, and add the appropriate correction, if the calibration data are to continue to be of use. This should be done by checking the reading of the thermometer at a convenient reference temperature at regular intervals. The reference temperature may be any temperature within the range of the thermometer that is reproducible to better than its calibration uncertainty. The most convenient and easily reproducible reference temperature is that of the ice point, which, if properly prepared, is reproducible to within 1 mK. For this reason, most good thermometers have a scale marked around 0°C. If the thermometer does not have such a scale, then another reference temperature must be chosen—usually the lowest temperature on the scale, although this will depend on the thermometer and on the resources available for carrying out the test.

Other fixed points may be used as reference temperatures, for example, the steam point (100°C). However, this is a more difficult point to realise because of variations in temperature, caused by changes in air pressure, and to internal pressure changes in the thermometer, resulting from the high temperature of the expansion chamber, so it is no longer a commonly used reference temperature. If the reference temperature is not a fixed point, the thermometer must be checked by comparison with a standard thermometer.

Unfortunately, the effect of temporary depression (section 5.3.2) on the reference-point reading must also be taken into account. Its effect on the reading opposes that of the secular change, so unless it is minimised or eliminated, it can be difficult to determine the full extent of secular change. Temporary depression depends on the individual thermometer and on the temperature to which the thermometer had been heated. As a means of minimising its effect, *thermometers are conditioned prior to measurements, at the reference point.*

6.2.3   Conditioning

Conditioning places a thermometer in a known and reproducible state, and is applied just before checking it at the reference temperature. If possible, the conditioning procedure should be chosen to minimise the temporary depression, although, as long as the magnitude of the temporary depression is the same each time the reference temperature is measured, the determination of secular change is still reliable. Conditioning procedures vary according to the range and accuracy of the thermometer. Those used at NML are described below and summarised in section 6.5.1.

Method I If the maximum test temperature is greater than 100°C, the thermometer is heated to this maximum and then cooled slowly over 12 to 15 h. This minimises the temporary depression, so very little change will occur in the thermometer until it is next heated. Consequently, some time can be allowed to elapse between the conditioning and the reference-temperature reading. In a new thermometer, secular change is also going to have a significant effect, so the conditioning procedure may have to be repeated a number of times until the reference-temperature reading is stable.

If slow cooling is not used the elapsed time between heating and reading the thermometer becomes important. For example, if a thermometer is heated, cooled quickly in air and then given an ice point check 1 h after it had been cooled, its reading may be significantly different from that obtained if checked 24 h later. The extra time will have allowed the bulb to contract a little more and the thermometer will read higher than it did the previous day. Consequently, if a reference point is checked after only 1 h at ambient and checked again six months later after 24 h at ambient, the two readings are **not** comparable and the secular change cannot be determined.

Method II The temporary depression arising from the difference between ambient temperature and the reference temperature can be significant for thermometers having a maximum temperature of use less than 100 °C and a required uncertainty less than 0.02 °C. For this reason such thermometers are kept below the reference temperature for a few days.

Method III Thermometers with a maximum temperature less than 100 °C and a very high accuracy requirement, i.e., an uncertainty of 0.01 °C or better, must be kept below 0 °C and the ice point monitored regularly until the reading is stable, i.e., the temporary depression had dissipated, before the secular change can be determined. This may take months.

Method IV The process described in method III can be shortened for some thermometers by heating the thermometer to its maximum temperature for $1/2$ h, cooling it and measuring the ice point immediately. A temporary depression component will be included but, if the reference-point reading is always done the same way, this component will be constant and any difference in the reading should be due to secular change.

Method V Thermometers having a maximum test temperature of less than 100 °C and a required uncertainty greater than 0.02 °C are stored at ambient temperature for at least one week before the reference point is read.

All checks at the reference temperature must be carried out with care. If a fixed point is not used, the thermometer is compared with a standard in the same way that a calibration is performed (see section 6.2). The ice point itself presents a few unique problems, partly because it must be properly prepared and used, if an accurate temperature is to be obtained, and partly because the thermometer reading is static.

The equipment and proper preparation of an ice point are detailed in section 8.2.3. The following procedure should be used to measure the ice point on a liquid-in-glass thermometer.

- The thermometer is inserted vertically into the ice.

- It should be supported at two, preferably three, points along its length to ensure verticality.

- Total-immersion thermometers should be immersed until the ice point reading is covered.

- Partial-immersion thermometers should be immersed to the set immersion and the emergent-stem temperature monitored, as described in section 6.2.6.

- The thermometers should be left in the ice for *at least* 5 minutes before being read.

- Before reading, a total-immersion thermometer is raised until a few divisions are visible above the ice.

- All thermometers are gently dumped before reading to avoid the effects of stiction. A thermometer is 'dumped' by lifting it a few millimetres above its reading position and then sharply but carefully dumping it back into position. Tapping may also achieve the desired result, but it sometimes has variable outcomes.

The thermometers should be read through a viewing system similar to that described in section 6.2.4.

6.2.4   Calibration equipment and set-up

To carry out a calibration of liquid-in-glass thermometers by comparison, the following items of equipment are the minimum required:

- a variable-temperature calibration enclosure,
- a calibrated temperature standard and any auxiliary equipment,
- a viewing system and
- a calculator or computer for data reduction.

The calibration enclosure provides a temperature-controlled region at uniform temperature, which places the standard and test thermometers at the same temperature. Stirred-liquid baths are the most commonly used because they can provide a large temperature-controlled volume, giving good thermal contact with the thermometer bulbs and are able to operate over a wide range of temperatures.

The spatial and temporal temperature variations in the bath should be measured before calibrations are conducted. This is done to ensure that any temperature variations in the working space are within acceptable limits. The limits will depend on the accuracy required of the thermometers to be calibrated in the bath. A general guide is that spatial variations over the working space should be less than half the required uncertainty. Information about calibration enclosures and their testing can be found in Chapter 8.

Liquid-in-glass thermometers are best calibrated with the temperature rising slowly, as this minimises the effect of stiction. For this reason, temperature control of the calibration enclosure should include a ramp function able to be set at a low rate. The optimum rate of rise depends on the relative response time of the test thermometers and the standard thermometer and the required calibration uncertainty—as a general rule, a rate equivalent to the expected uncertainty per minute is usually low enough. Ideally, then, the ramp function should be adjustable to suit a range of thermometers. If this is not possible, the rate of rise should be set at a level suitable for the most accurate thermometers likely to be calibrated.

Any calibrated thermometer (PRT, thermocouple, liquid-in-glass, etc), with a suitable temperature range and accuracy, can be used as the standard for the calibration. The uncertainty in the temperature measured by the standard, including contributions from any auxiliary measuring equipment (DVM etc, if required), should be less than half, and preferably less than one fifth, of that required of the thermometer being calibrated, so that its contribution to the total uncertainty is not significant. The most commonly used standards are liquid-in-glass and platinum-resistance thermometers. At NML, SPRT's are used for nearly all calibrations, except those done at temperatures below $-50°C$, when a platinum/rhodium resistance thermometer is used.

If liquid-in-glass thermometers are used as standards, it is good practice to use two of them at each test temperature. One advantage of this is that they act as a check on each other. Each standard should have a current calibration and a recent reference-point check, so that any secular change correction can be added. For calibrations at the highest accuracy, this check should be carried out just prior to use. The appropriate conditioning procedures should be carried out on the standards as well as the thermometers being tested.

Liquid-in-glass thermometers cannot be read accurately with the naked eye. A properly constructed viewing system is required to see the top of the meniscus clearly, to magnify the divisions sufficiently for accurate interpolation and to avoid parallax problems. With an experienced observer and a properly set up viewing system (with a graticule, if necessary), a thermometer may be read reliably to better than $1/10$ division.

Although small lenses can be obtained that attach to the thermometer itself, these are not suitable for use in calibration as there is quite a bit of distortion around the edges, they are restricted to one thermometer and they do not necessarily eliminate parallax errors. The most suitable equipment is a telescope with rack-and-pinion focussing, mounted on a stand that provides for vertical movement and horizontal rotation of the telescope. It should have a magnification of 5 to 10, a focal length of 30 to 40 cm and a depth of focus of at least 1 cm. At NML a binocular telescope is used for most readings, but a monocular telescope is equally suitable. Note that the image through some telescopes is inverted.

The magnification is chosen to give a clear image of the meniscus and a substantial portion of the scale without enlarging the scale so much that it is difficult to interpolate between divisions. A relatively small focal length allows the reader to reach the thermometer while still looking through the telescope. This is important if dumping is necessary or if the thermometers need to be rotated for better vision. If the telescope is too close, the visible portion of the scale is greatly restricted and the observer may mistake the section of scale being read. Depth of focus is important as the scale and the meniscus need to be in focus simultaneously, and there can be as much as 6 mm between them in a solid-stem thermometer.

Parallax is the apparent displacement of an object against a background as the angle of view changes. The distance between the meniscus and the scale on a solid-stem thermometer can result in substantial parallax errors if not viewed straight on (at right angles).

Regardless of the type of standard thermometer used, calibration by comparison requires some data manipulation. This is most easily done with the aid of a computer, which can also store all necessary information and print up the report. A calculator can do the calculations very quickly, but they can also be done easily by hand, if necessary, except for the conversion of resistance to temperature.

Initial set-up

The thermometers to be calibrated and the standard thermometer(s) are placed in the temperature-controlled bath. It is simpler to place them in a straight line across the bath, as shown in Figure 6.1. Although, as long as they are within the tested working space of the bath they can be arranged in any way that will allow them to be read in a repeatable sequence. Total-immersion thermometers are immersed so that just a few divisions around the test point are visible, while partial-immersion thermometers are immersed to the immersion line. Ideally, the bulb of the thermometers should be at the same depth as the sensor of the standard, but this is rarely practical due to the variable immersion of liquid-in-glass thermometers, so it is often necessary to rely on the uniformity of the bath. Information on measuring the uniformity of calibration enclosures is given in Chapter 8.

If only one thermometer is to be calibrated at a particular temperature, then the calibration method can be simplified, as readings need not be taken across a line of thermometers and back. As long as the rate of rise is low enough, a reading of the test thermometer can be immediately followed by a reading of the standard thermometer. One advantage of this is that the

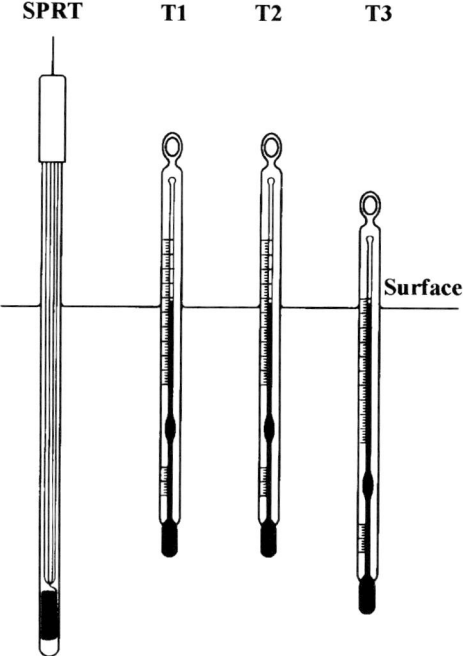

Figure 6.1: Arrangement of thermometers for calibration.

test thermometer can be read when the meniscus is at a graduation line, so interpolation is not necessary. This method is *not* suitable when using liquid-in-glass thermometers as standards.

The bath is set to a temperature just a little below the chosen calibration temperature. When this temperature is reached, and the thermometers are in thermal equilibrium with the bath, the calibration can begin.

6.2.5 Calibration procedure for total-immersion thermometers

The thermometers are inspected through the telescope to ensure that all are visible and the readings are within five divisions of the chosen calibration temperature. A thermometer is easier to read if rotated until the liquid is lined up along one side of the scale. Any necessary alterations to position are made and then the temperature ramp is activated at the appropriate rate for the most sensitive thermometer being calibrated. When the temperature is rising at a constant rate the readings are begun.

The readings are taken from left to right and then immediately repeated right to left. So the SPRT (of Figure 6.1) is read first, then T1, T2, T3, T3, T2, T1 and finally the SPRT reading is repeated. This technique relies on the rate of temperature rise being constant and the time between readings being substantially the same. If this is so, then the average of each pair of readings— effectively correcting all data to the same time—is equivalent to simultaneous readings of all the thermometers. If they were all read at the same time then they were all at the same temperature and the test thermometers can be legitimately compared with the standard.

When reading a number of thermometers in this manner, it is not possible to have the meniscus of each thermometer conveniently at a scale marking, so interpolation is necessary. Interpolations should be made between the middle of one scale line and the middle of the next, not between the top of one marking and the bottom of the next. On thermometers with very fine scale markings, this will not make much difference, but it can make a significant difference if the scale markings are thick.

After the set of readings is taken, the average of each thermometer reading is determined and noted. The calculation of the bath temperature from the average resistance reading and the determination of the corrections can be done either at the end of the set or when all sets of readings at the test temperature are complete. At least three sets of readings are taken at each test temperature and a typical calibration set is shown in Table 6.1. Readings are taken above and below the point being tested. This will cause any variation in the scale on either side of the point to be included in the scatter of the readings and thus in the assessment of the uncertainty. Three readings at each point are the minimum that will give any meaningful statistical information.

Most observers have an individual bias to their readings, due to individual interpretations of what is seen. If possible the readings at each point should be taken by two observers to minimise the effect. This is especially important if one of the readers is inexperienced. Practice will reduce the bias, but it is impossible to eliminate it entirely.

Once the readings are completed and the corrections calculated, the corrections are averaged and their scatter is determined. If the results are satisfactory, i.e., the thermometer had shown no unusual characteristics, the correction is reasonable for that type of thermometer and the scatter of corrections is within acceptable limits, then the bath temperature is set just below the next test point and the whole process is repeated until all test points are complete. If there is a problem at any test point, then an effort must be made to determine the cause. If necessary, the point may have to be repeated.

Then, the thermometer is conditioned again and the reference temperature is checked to give a measure of any residual temporary depression and an

Table 6.1: An example set of calibration results for thermometers T1, T2 and T3, and the calculation of corrections.

Readings		SPRT (Ω)	Thermometer Readings (°C) T1	T2	T3	Temperature[†] (°C)
1st set		28.1430	19.770	19.760	19.781	
		28.1440	19.784	19.763	19.781	
	Average:	28.1435	19.777	19.762	19.781	19.704
	Correction:		−0.072	−0.057	−0.077	
2nd set		28.1560	19.892	19.885	19.903	
		28.1568	19.898	19.887	19.903	
	Average:	28.1564	19.895	19.886	19.903	19.829
	Correction:		−0.066	−0.057	−0.074	
3rd set		28.1760	20.092	20.078	20.100	
		28.1770	20.098	20.080	20.100	
	Average:	28.1765	20.095	20.079	20.100	20.0234
	Correction:		−0.072	−0.056	−0.077	
Result	Av. correction:		−0.070	−0.057	−0.076	
	Scatter:		0.006	0.001	0.003	

[†] values calculated from average resistance of SPRT.

indication of the repeatability of the reference temperature. The difference should be considered when calculating the uncertainty of calibration.

The collected information is summarised. This makes it easy to check for anomalies in the corrections or the scatter. If the results are satisfactory, the report is printed. A possible layout of a result page is shown in Figure 6.2. Note that this is not the full report, which should also include a summary of the test method and any notes.

6.2.6 Calibration procedure for partial-immersion thermometers

The procedure for calibrating partial-immersion thermometers is very similar to that for total-immersion thermometers. The main difference is that the emergent-stem temperature must be monitored during calibration. This is commonly done by using a Faden thermometer or a number of short-stem thermometers.

The Faden or stem temperatures are recorded with the readings and the emergent-stem temperature is calculated for each calibration point (see below), as shown in Table 6.2. The calculated emergent-stem temperature for each calibration point should be included as a fourth column in the results table

Measurement report on Mercury -in-glass Thermometer:
Serial No: T1

Range: −1 to 51
99 to 101 °C

Graduation: 0.01 °C   **Uncertainty:** 0.02 °C

Immersion: To the reading, vertical

Reference Temperature: 0.0 °C

Conditioning Procedure: Before testing, the thermometer was kept below its reference temperature for at least 24 hours.

Reading	Correction	Temperature
0.000 °C	+0.000 °C	0.000 °C
5.000	−0.065	4.935
10.000	−0.055	9.945
15.000	−0.075	14.925
20.000	−0.070	19.930
25.000	−0.060	24.940
30.000	−0.050	29.950
35.000	−0.070	34.930
40.000	−0.035	39.965
45.000	−0.035	44.964
50.000	−0.025	49.975
−0.005	+0.005	0.000

No specification test was done on this thermometer.

Notes:/

Reference No: File Number: Checked: Date:

Figure 6.2: Sample NML report for a total-immersion thermometer.

Table 6.2: Typical calibration data for a partial-immersion thermometer, T4, when using a platinum-resistance thermometer (SPRT1) as the standard and a Faden thermometer (F1) to indicate the stem temperature. The Faden bulb was 56 mm long and was immersed 10 mm.

SPRT1 (Ω)	Thermometer T4 (°C)	Faden F1 (°C)	Temperature[†] (°C)	Calculated Correction (°C)
33.4883	79.900	45.8	79.908	+0.008
33.5084	80.100		80.111	+0.011
33.5186	80.200	46.4	80.214	+0.014
			Average:	+0.011
	Average:	46.1	Scatter:	0.005
Calculated[‡] emergent-stem temp. = 38.1				

[†] calculated from SPRT1.   [‡] from equation (6.1).

of the report (compare results given in Figure 6.2). Calibration of a partial-immersion thermometer reported without the emergent-stem temperature is virtually useless, as there is then no way of determining whether the emergent-stem temperature differs from that in use.

Obviously, the emergent-stem temperature must also be monitored during use and appropriate corrections made (the calculation of emergent-stem corrections is discussed in section 6.4.1).

Using a Faden thermometer

A Faden thermometer has a long bulb designed to simulate the emergent stem of the test thermometer. If used correctly, it not only responds to ambient temperatures in much the same way as the column of the test thermometer, but it also includes a conduction component.

The Faden thermometer is placed in the bath so that the top of its bulb is at the same level as the meniscus of the test thermometer and with at least 10 mm of bulb immersed in the bath. Faden thermometers are available in a number of bulb lengths, but it is not necessary to obtain a whole set, as the length of bulb exposed can be varied by changing its immersion to any depth provided it is at least 10 mm. The Faden thermometer should also be placed as close as possible to the thermometer to which it refers so that the surrounding temperature can be assumed to be identical for both thermometers. Figure 6.3(a) shows the way a Faden thermometer can be used to measure the mean temperature of the emergent-liquid column.

The advantage of a Faden thermometer is not only that it reproduces more accurately the conditions affecting the emergent column, but it usually requires the reading of only one instrument to obtain the information from which the emergent-stem temperature is calculated (unlike the use of stem-thermometers). The average emergent-stem temperature, t_{es}, is calculated as follows:

$$t_{es} = \frac{l_t t_f - l_i t_b}{l_t - l_i}, \qquad (6.1)$$

where l_t is the total length of the Faden bulb, l_i is the length of bulb immersed in the bath, t_f is the reading of the Faden thermometer and t_b is the temperature of the bath.

Using stem-thermometers

A 'stem-thermometer' has a short stem and a small bulb and a number of them are placed along the length of the emergent stem to monitor the surrounding air temperature. Each stem-thermometer is assumed to indicate the temperature of the section of emergent column adjacent to it. The number used will depend on the length of this column. In regions where the temperature is likely to be very non-uniform, a stem-thermometer would represent only a small length of the column, while in regions where there is not likely to be much variation, a single thermometer represents a relatively long length of emergent stem. Figure 6.3(b) shows the arrangement of stem-thermometers along a column of moderate length.

The middle of the bulb should be placed at the centre of the length of emergent stem whose temperature it is monitoring. The one closest to the bath should cover 30 to 50 mm and the others, up to 100 mm each. The average emergent-stem temperature can be calculated from the relationship:

$$t_{es} = \frac{d_1 t_1 + d_2 t_2}{d_1 + d_2},$$

where d_1 and d_2 are the lengths covered by stem-thermometers 1 and 2, respectively, and t_1 and t_2 are their readings. If a third stem-thermometer is required then the additional values, d_3 and t_3, must be included in the above equation ($d_3 t_3$ in the numerator and d_3 in the denominator).

The lengths d_1 and d_2 can be expressed in normal length units, such as millimetres, but will then not reflect changes in bore diameter or in the expansion coefficient. Thus, the lengths are best expressed as 'equivalent degrees'. When the scale does not extend to the immersion line, the degree equivalent of the ungraduated portion must be determined by measuring the number of degrees in the same length of adjacent scale.

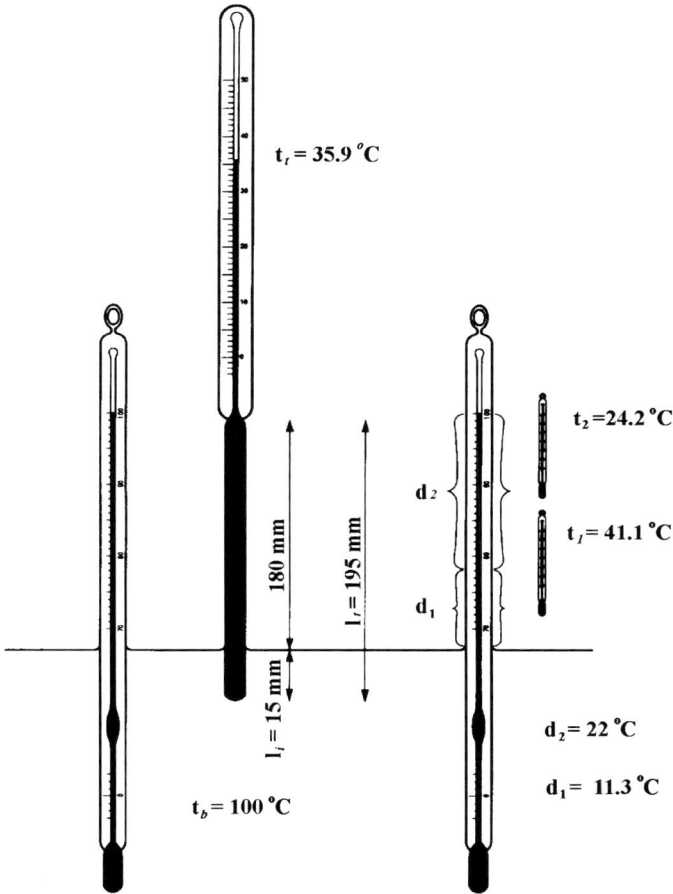

Figure 6.3: The measurement of emergent-stem temperature by means of (a) a Faden thermometer and (b) using short-stem thermometers.

If the thermometer being tested has a contraction or an enlargement of the bore in the exposed part of the column, then the volume of liquid in that space must be calculated or measured, the equivalent number of degrees calculated and its temperature measured using a separate stem-thermometer. With the exception of some specialised thermometers, well-designed, partial-immersion thermometers will not have enlargements in the exposed parts of the column.

6.2.7   Calibration using liquid-in-glass standards

When liquid-in-glass thermometers are used as standards, the calibration procedure is similar to that described in section 6.2.5, with some slight variations and complications. As previously mentioned (section 6.2.4), two liquid-in-glass thermometer standards should be used. Their calibration uncertainties should be suitably small and they should be prepared and used so that their best accuracy is achievable. In other words, they should be properly conditioned, reference-point checks should be carried out and their respective secular changes calculated.

The thermometers are arranged in the bath as before, but with one standard at each end of the line. As before, the thermometers are read from left to right and back again with the bath temperature rising slowly. The average reading is determined for each thermometer and the standards are corrected for secular change and calibration error. After correction, the readings of the two standards should agree to within their combined uncertainty. The bath temperature is taken to be the average of the corrected readings of the two standards. A sample set of readings for this type of calibration is given in Table 6.3.

6.3   Uncertainty of calibration

Every measurement involves a degree of uncertainty, and if a measurement result is stated, its uncertainty should be included along with its confidence level, i.e., the probability that the true value lies within the stated range of uncertainty. NML follows the ISO Guide [21, 22] in calculating and expressing uncertainties—it expresses the calibration uncertainty as an 'expanded uncertainty', effectively the confidence limits, and does so at a confidence level of 95%.

Although most components of uncertainty need to be estimated, the component is not a number plucked out of the air. When a measurement is carried out, all of the possible sources of error must be identified, evaluated and combined to give a figure for the total uncertainty. Examples of uncertainty contributions that are known are the calibration uncertainty of the standard and the measured spatial variation of temperature in the calibration enclosure. Further information on the management and expression of errors is given in references [21, 22].

Some of the common sources of uncertainty in the calibration process are:

- the standard thermometer(s),
- the equipment used to read the standard e.g the resistance bridge,

Table 6.3: Example of calibration results when using liquid-in-glass standards (S1 and S2).

	Standard S1 (°C)	Thermometer Readings (°C) T5	T6	Standard S2 (°C)	Average S1 & S2 (°C)
Readings:	4.928	4.975	4.960	4.949	
	4.931	4.980	4.960	4.949	
Average:	4.930	4.978	4.960	4.949	
Correction to Std:	+0.026			+0.005	
Temperature:	4.956			4.954	4.955
Correction:		−0.023	−0.005		
Readings:	4.974	5.020	5.015	4.994	
	4.978	5.020	5.018	4.994	
Average:	4.976	5.020	5.017	4.994	
Correction to Std:	+0.027			+0.005	
Temperature:	5.003			4.999	5.001
Correction:		−0.019	−0.016		
Readings:	4.990	5.025	5.025	5.007	
	4.994	5.030	5.030	5.007	
Average:	4.992	5.028	5.028	5.007	
Correction to Std:	+0.027			+0.005	
Temperature:	5.019			5.012	5.016
Correction:		−0.012	−0.012		
Av. Correction:		−0.018	−0.011		
Scatter:		0.011	0.011		

- measurement of the emergent-stem temperature,
- temperature variations in the calibration enclosure,
- limit of reading of the test thermometer,
- secular change,
- temporary depression,
- changes in ambient temperature or pressure,
- stiction and
- capillary irregularities.

Each of these factors should be assessed and combined to give the total uncertainty. Many of the factors remain the same for each calibration or for a particular calibration range, so it is often convenient to determine the least possible uncertainty for a thermometer of a particular range and graduation size, and to assess from the examination and calibration of individual thermometers whether any factor is excessive and thus calls for an increased uncertainty. For instance, the uncertainty of the standard and the measuring equipment would be constant over a wide temperature range; the uncertainty due to spatial variations in the bath might be the same for the limited temperature range covered by that bath; the ambient pressure and temperature changes are usually small enough to ignore, but would be considered for particularly sensitive thermometers; the effect of stiction is usually included in the uncertainty of the reading.

The set of 'least uncertainties' used at NML is given in Table 6.5 in the appendix. Some laboratories may be able to use these as a basis for their own calculations, but it should be remembered that there are underlying assumptions about the equipment used (see section 6.5.2), and that if these are not valid for a particular laboratory the tabulated values should be modified or re-determined. For example, one of the assumptions is that the spatial temperature variation in the comparison bath is $\leq 0.002\,^\circ$C for the range 0 to 250 $^\circ$C. If the spatial variation in the calibration enclosure used is 0.02 $^\circ$C then obviously the calibration uncertainties would have to be revised significantly.

6.4   Using a calibrated thermometer

A calibrated liquid-in-glass thermometer is a valuable instrument, with much of the investment in the calibration itself. So, it is important to try and get the best use out of the instrument, by caring for it properly and by using the calibration report correctly. It is not always a simple matter of putting the thermometer in the liquid and taking a reading.

6.4.1   The calibration report

A calibration report contains information that is general and some that is specific to the particular thermometer referred to in the report, so it should be read carefully. There can be a great deal of variation in the way a report is presented, though certain information is vital. A report is often more than one page, especially if there are notes. A report should contain:

- the thermometer serial number,
- the report reference number,
- the file number,
- the report date,
- a description of the calibration and conditioning methods,
- the uncertainty of the calibration, adequately expressed,
- the depth of immersion of the thermometer and
- the calibration table/results.

The table of results shown in Figure 6.2 lists the corrections to be applied to the thermometer when it is indicating the temperature in the column marked 'Reading'. The true temperature is then found by adding the correction to the reading, and this result is given in the third column headed 'Temperature'. If the thermometer had been calibrated at partial immersion, then there will be a fourth column that lists the emergent-stem temperature at each calibration point. There are other possible ways of laying out the table.

The thermometer had been calibrated at a limited number of points in its range, and measurements taken with it often lie between calibration points. When this happens, the appropriate correction is determined by linear interpolation. The calibration can be used to prepare an interpolation graph of corrections against temperature. Note that it is *not* valid to use a least-squares fitting of the corrections to a polynomial.

Since secular change should be monitored throughout the active life of a thermometer, at a suitable reference temperature, the calibration correction at this reference temperature may be subtracted from the reported corrections and a current reference-temperature correction added later at the time of use. This is equivalent to adding the secular change to the calibration. Consequently, it is convenient to prepare the interpolation graph with the reference-temperature correction subtracted. Figure 6.4 shows an interpolation graph, with the ice-point correction subtracted.

For partial-immersion thermometers, the report also lists the emergent-stem temperature for each calibration point. This is used with the emergent-stem temperature measured during use to calculate the corresponding correction Δ, using the equation already given in Chapter 5:

$$\Delta = Kn\,\delta t$$

where K is the differential coefficient of expansion for the liquid in glass, n is the number of graduations exposed (in equiv. °C) and δt is the difference between the emergent-stem temperature in calibration and that in use—its sign being critical: $\delta t = t_{cal} - t_{use}$. For mercury, $K = 0.00016°\text{C}^{-1}$, although it varies slightly with temperature and glass type. The variation is greater for spirit thermometers, but for the most common ranges, $K = 0.00096°\text{C}^{-1}$ applies.

If the thermometer is graduated all the way down to the immersion line, then n can be easily determined from the scale. Otherwise, the length of the exposed, ungraduated stem must be measured and the equivalent number of degrees determined by comparing it with the same length of graduated stem.

Consider the example of a partial-immersion thermometer being used at 250°C. Assume that it has an exposed-column length equivalent to 200°C and

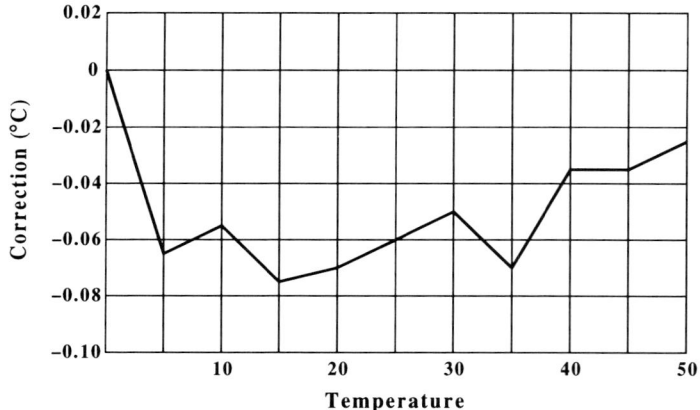

Figure 6.4: Interpolation graph for thermometer T1, part of whose calibration report appears in Figure 6.2.

that its emergent-stem temperature in use is 40°C. If the calibration report gives its emergent-stem temperature during calibration as 30°C, then:

$$\Delta = 0.00016 \times 200 \times (30 - 40) = -0.32°C.$$

This correction should be added to the calibration correction in the report.

If a total-immersion thermometer is used at partial immersion, the same equation can be used. Say, for instance, that the total-immersion thermometer in Figure 6.2 was used at an immersion depth of 100 mm and gave a reading of 50°C, leaving an exposed column 75 mm long. If the emergent-stem temperature was 25°C, rather than 50°C (the bath temperature), as it was during calibration, then:

$$\Delta = 0.00016 \times 72 \times (50 - 25) = 0.290°C.$$

Total-immersion thermometers used at partial immersion will often have large corrections, especially at very high temperatures where corrections could be ~20 to 30°C.

The report should indicate the conditioning method used for calibration, and this should be used, if possible, for later reference-temperature checks. Additional notes will specify any further corrections that might be required, any restrictions on the calibration or any changes in the conditions of use.

6.4.2   General care of the thermometer

A calibrated thermometer should have been examined by the calibrating authority before being tested, so it should have no unrectifiable faults. However, it should still be examined carefully for column separation and other repairable faults when it is received, as such faults are often induced by transporting the thermometers. Most liquid-in-glass thermometers can be stored flat, but should be kept in slotted trays to reduce movement.

Before use, the thermometer should be conditioned, the reference temperature checked and the reading compared with that in the calibration report. If there is a significant difference, say half the uncertainty, then possible causes of this difference should be considered. If the reading is high, check again for gas bubbles in the bulb or a separated column; if it is low, check for liquid in the expansion chamber and the conditioning method used.

The reference temperature should be checked frequently when the thermometer is new and at intervals of about 6 months once it becomes stable. Although there will be a slight rise in the reading over time due to secular change, any significant variation should be investigated.

If a thermometer is being used as a standard for the calibration of other thermometers, then a reference-temperature reading should be taken before each use.

Repairing faults

Separated column. To fix a separated column, the liquid, usually mercury, must be drawn back into the bulb, where the separated section can be rejoined by gentle tapping. This is done by immersing the bulb in crushed dry ice or, very carefully, cooled over liquid nitrogen. No harm results if the mercury freezes as, unlike water, it contracts on freezing. However, care should be taken when warming the bulb. If the mercury at the bottom melts first and is trapped by frozen mercury above, then the pressure exerted can break the bulb. For this reason, the thermometer bulb is usually warmed gently from the top.

Gas bubble in bulb. The bubble is moved to the top of the bulb, preferably right under the capillary, by keeping the thermometer upright and gently tapping. After this, the process used for fixing a separated column is used. If the gas bubble is very small, it may not move and it may then be necessary to deliberately produce an extra bubble in the bulb. This bubble can then be moved around until it absorbs the smaller one. Then an attempt should be made to remove the large bubble in the usual way, although, because of the size of the bubble, it may not be possible to contract all the mercury into the bulb before it freezes.

Mercury droplets in capillary. If the droplets are discrete and do not block the capillary, then they can be picked up by heating the thermometer until the mercury column reaches them. This should *not* be done if the droplets are above the maximum scale mark. In this case, the column should be deliberately separated so that the mercury can be taken up above the upper scale reading without actually heating the thermometer above this temperature. Of course, then the separated column must be mended.

Broken-spirit column. Sometimes the spirit in a broken column will rejoin of its own accord, if stored upright and allowed a period of a few days to drain. If this does not work, it may be necessary to heat the top of the thermometer to evaporate the broken section. To make sure that it condenses back in the main column, the bottom part of the thermometer should be kept cool. Great care must be taken as the thermal stress can break the thermometer.

Faint scale lines. If the lines are etched into the glass, then improving faint lines is a simple matter of replacing pigment that had come out. This can be done by rubbing over with a paste-like mixture of sodium silicate, water and a black oxide, such as MnO_2, and then wiping the excess mixture off. Care should be taken though, because if the mixture dries on the thermometer it can bond to the rest of the glass and make the thermometer even more difficult to read.

Most other repairable faults are fixed using a combination of the methods already described.

Etching

There are some occasions when it is necessary to etch information onto a thermometer, usually a serial number. Thermometers that are made to a recognised specification are required to have a unique serial number, so no etching should be required in such cases. However, other thermometers often lack a serial number, so, if they are calibrated, a serial number must be permanently etched onto them. This is best done using hydrofluoric acid. The thermometer is first coated with wax or nail-polish and the number is engraved in this coating using a metal stylus and a pantograph. A wall of petroleum jelly is then built up around the engraved section and a few drops of hydrofluoric acid are dropped inside this dam. After five to ten minutes, depending on the strength of the acid and on the glass type, the thermometer is carefully rinsed and cleaned. The number is blackened using the method described above for repairing faint lines. **WARNING:** Hydrofluoric acid is extremely dangerous, both on skin contact and inhalation. Any procedure using it should be carried out under a fume hood and using rubber gloves.

6.4.3 Transport

Unfortunately, it is sometimes necessary to transport a thermometer—to a calibration laboratory, to a customer or to a new location. It is unfortunate because liquid-in-glass thermometers are very fragile and are often broken during relocation. If possible, the thermometer should be carefully packed and transported by hand. If this is not possible, it should be packed in a rigid tube, but buffered and supported along its whole length by soft material, so that it does not move and strike the sides, top or bottom of the container. This tube should then be packed tightly inside another container. Care should be taken that the fit not be so tight that the thermometer is broken by the packing process.

6.5 Appendix

6.5.1 Conditioning procedures in summary

A suitable conditioning method may be selected from Table 6.4. The various procedures are discussed in section 6.2.3, beginning on page 110.

Table 6.4: Selection of conditioning method to suit the maximum test temperature, t_m, and desired calibration uncertainty, U.

Uncertainty (°C)	Maximum Test Temperature		
	$t_m > 100°C$	$t_m \leq 100°C$	$50 \leq t_m \leq 100°C$
$U \geq 0.05$	Method I	Method V	Method V
$0.01 < U \leq 0.05$	Method I	Method II	Method II
$U \leq 0.01$	Method I	Method III	Method IV

Below is a step by step summary of the conditioning procedures used at NML prior to calibration. After calibration, the secular change should be checked by the user (section 6.4.2), but before doing so the following conditioning steps marked (#) should be followed.

Method I

- Heat to the maximum test temperature for a few hours (#),
- cool slowly over at least 12 h (#),
- measure the reference-temperature correction (#),
- repeat the above steps until the ice-point correction is constant,

- CALIBRATE,
- cool slowly from the maximum test temperature,
- measure the reference-temperature correction and
- USE – but cool slowly after each use.

Method II

- Store at or below the reference temperature for at least 24 h (#),
- measure the reference-temperature correction (#),
- CALIBRATE,
- store below the reference temperature overnight and
- measure the reference-temperature correction.

Method III

- Store below the reference temperature, at least overnight (#),
- measure the reference-temperature correction (#),
- repeat the previous two steps over a few weeks until the reference-temperature correction is constant,
- CALIBRATE,
- store below the reference temperature overnight and
- measure the reference-temperature correction.

Method IV

- Heat to the maximum test temperature for about $1/2$ h (#),
- store at or below 0°C for approx 16 h (#),
- measure the ice-point correction (#),
- CALIBRATE,
- store at or below 0°C for the same time period as above and
- measure the ice-point correction.

Method V

- Store at ambient temperature for at least one week (#),
- measure the reference-temperature correction (#),
- CALIBRATE,
- store at ambient temperature for about 16 h and
- measure the reference-temperature correction.

6.5.2   Least uncertainties of calibration

The least uncertainty to be expected from a liquid-in-glass thermometer calibrated at NML is given in Table 6.5. The values are based on large numbers of thermometers and do not necessarily apply to any individual instrument. They refer to well-made thermometers calibrated with the following conditions/assumptions applying.

- The spatial temperature variation in the calibration bath is $\leq 0.002°C$ up to 250°C and 0.1°C above 250°C

- The combined uncertainty of the standard and its measuring equipment is <0.002°C

- The thermometer can be read accurately to 1/10 of the smallest graduation interval.

- the standard deviation of the readings at any calibration temperature is less than half the overall uncertainty.

- The thermometer is kept at the calibration temperature long enough to reduce the effect of any further expansion of the glass to about 0.002°C

- The thermometer is used at the correct immersion and under normal ambient conditions.

- The readings are taken with the temperature rising except at the ice point.

- Proper conditioning procedures are followed.

- The number of calibration points is sufficient to avoid significant interpolation error.

- There are no visible anomalies in the bore or the scale.

- The thermometer is used under normal atmospheric pressure.

The data in Table 6.5 are used as follows. If the conditions applying to a thermometer and to its calibration satisfy those listed above, the calibration uncertainty would be the relevant value in the table. Otherwise, the value is adjusted by allowing for any significant departures from the listed conditions.

Table **6.5:** The minimum calibration uncertainties for totally-immersed (upper values in table) and partially-immersed thermometers (lower), given as a function of smallest scale division and temperature range. The lower values apply only if the immersion exceeds 57 mm* for mercury-filled thermometers and 76 mm* for spirit.

Smallest Division (°C)	Min. Uncertainty for Temperature Range (°C):							
	−183 to 50	−80 to 50	−50 to 50	−10 to 50	−10 to 100	−10 to 200	−10 to 260	−10 to 450
0.01				0.005	0.01			
				0.005	0.01			
0.02			0.02	0.01	0.02			
			0.02	0.01	0.02			
0.05			0.05	0.01	0.02	0.05		
			0.05	0.02	0.05	0.05		
0.1			0.1	0.02	0.02	0.05	0.1	
			0.1	0.02	0.05	0.1	0.2	
0.2			0.1	0.05	0.05	0.05	0.1	
			0.1	0.05	0.1	0.1	0.2	
0.5	0.5	0.5	0.2	0.1	0.1	0.1	0.2	0.2
	0.5	0.5	0.5	0.2	0.2	0.2	0.2	0.5
1	1	1	0.2	0.2	0.2	0.2	0.5	0.5
	1	1	0.5	0.2	0.2	0.5	0.5	0.5
2	2	1	0.5	0.5	0.5	0.5	0.5	1
	2	1	1	0.5	0.5	0.5	1	1

* If the immersion is less than this the uncertainty will be higher.

6.5.3 Some common thermometer specifications

This list is not exhaustive. A specification may contain many different designations. Copies of Australian Standards and British Standards can be obtained from Standards Australia.

Table 6.6: Examples of thermometer specifications.

Specification	Designation	Immersion (mm)	Range[†] (°C)	Graduation (°C)
BS593	A10C/total	total	-30 to 110	0.1
	A20C/total	total	-20 to +20	0.1
	A40C/total	total	0 to 40	0.1
	B60C/total	total	-20 to +60	0.2
	A100C/100	100mm	ip, 70 to 100	0.1
	F175C/100	100mm	ip, 123 to 177	0.5
AS2831	AS2831/0.1/-25/+5	total	-25 to +5	0.1
	AS2831/0.2/35/85	total	ip, 35 to 85	0.2
	AS2831/0.2/155/205	total	ip, 155 to 205	0.2
	AS2831/0.5/45/105	total	ip, 45 to 105	0.5
	AS2831/1/180/420	total	ip,180 to 420	1
ASTM	3C	76mm	-5 to 400	1
	113C	total	-1 to 175	0.5
	124C	total	-25 to -15	0.1
IP	30C	total	ip, 23.6 to 26.4	0.05
	74C	61mm	-35 to 70	0.5
	98C	44mm	100 to 300	2

† ip – ice point.

Chapter 7

Special Types of Expansion Thermometer

Corinna Horrigan

7.1 Liquid-in-glass thermometers

Special liquid-in-glass thermometers have been developed for specific applications, and such instruments often require extra care in use and special calibration techniques. This chapter looks at some of these thermometers, to explain how they work and what gives them their special characteristics, and briefly discusses any special techniques required in calibration or use. In many cases, electrical thermometers, such as PRT's, thermocouples and thermistors, may be used in their place.

7.1.1 Bomb-calorimeter thermometers

In bomb calorimetry, the heat given out by the combustion of a material is measured by observing the temperature change it causes in a known volume of water. The change is quite small, usually only a few degrees, but it must be measured with great precision. Special high-sensitivity thermometers, known as bomb-calorimeter thermometers (see Figure 7.1), have been developed to do this. They have a very small range, typically 6 °C, and their scales are very finely divided—usually to 0.01 °C. After calibration, and if used correctly, they can measure temperature differences to an uncertainty of 0.005 °C or better.

High sensitivity is achieved by combining a large bulb with a very fine bore, but this very sensitivity allows some effects, usually negligible in other types of thermometer, to become important. For example, changes in air pressure can significantly affect the reading. In use, this is rarely a problem,

135

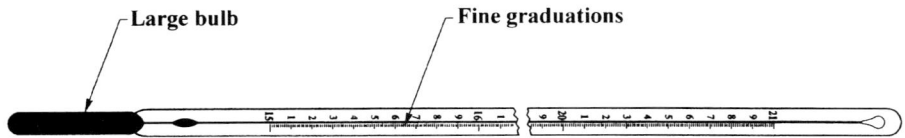

Figure 7.1: Schematic diagram of a bomb-calorimeter thermometer.

as only temperature differences are required and the effect is the same at all readings. However, it can cause errors during calibration, because significant atmospheric changes may occur, especially if the calibration takes more than one day to complete. For this reason, it is wise to monitor the air pressure during calibration.

Of greater significance, both in calibration and in use, is the internal pressure change that is caused by a change in the temperature of the expansion chamber. These thermometers are gas-filled and an increase in the temperature of the gas will increase the internal pressure and depress the reading. Consequently, if the temperature of the expansion chamber during use differs significantly from its temperature during calibration then an extra correction must be applied to the reading. The magnitude of the effect depends on the amount of gas in the thermometer and the type of glass used in its construction.

Table 7.1: A sample determination of the sensitivity of a bomb-calorimeter thermometer to changes in temperature of its expansion chamber.

Bath Temperature (°C)	Thermometer Reading (°C)	Correction (°C)	Expansion Chamber Temperature (°C)
30.000	30.010	−0.010	23.5
29.993	30.010	−0.017	0.0
	Differences:	+0.007	23.5

The sensitivity to expansion-chamber temperature is: $0.007/23.5 = 0.0003\,°C/°C$.

A bomb-calorimeter thermometer is first calibrated using the method described in Chapter 6 with the expansion chamber at ambient temperature. Then, the sensitivity to changes in temperature of its expansion chamber is determined for the highest scale reading obtained during calibration. At NML, this is done by lowering the temperature of the expansion chamber to 0°C, by surrounding it with ice, and re-calibrating the thermometer at the chosen scale reading (see Table 7.1)—achieved by cooling the bath a little and re-ramping its temperature.

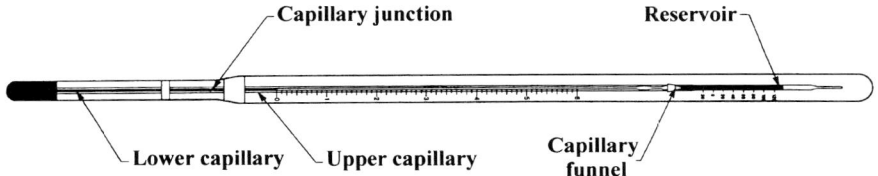

Figure 7.2: Schematic diagram of a Beckmann thermometer.

The fine bore can also cause problems with stiction, and it is possible for the meniscus to jump more than half a division. For this reason the thermometer should be observed carefully during calibration and use. If stiction is noticeable then the thermometer must be dumped (section 6.2.3) or tapped gently before each reading.

While these thermometers are usually marked for total immersion, they are rarely used that way. At NML the thermometers are calibrated at the immersion depth at which they are to be used. This ensures that the user is aware of the need for emergent-stem corrections. Calibration at partial immersion adds to the uncertainty of calibration, but a similar uncertainty would be incurred by using a thermometer, calibrated at total immersion, at partial immersion.

7.1.2 Beckmann thermometers

The Beckmann thermometer, like the bomb-calorimeter thermometer, is designed to measure small temperature changes accurately and usually has a scale range of only 5 or 6 °C. The lowest major scale marking is usually labelled zero. However, the thermometers are specially constructed to allow mercury to be added to or removed from the bulb, thus altering the temperature, the *setting*, at which the thermometer indicates 0 °C. Both the calibration and the correct use of a Beckmann thermometer are more complex than those of a bomb-calorimeter thermometer.

The thermometer (see Figure 7.2) must be vacuous to allow the mercury to be moved easily from the bulb to the reservoir, above the scale where the excess mercury is stored. A fine needle capillary allows very small drops of mercury to be removed from the main section of the thermometer into the reservoir, permitting the instrument to be set quite accurately—with a little patience.

Usually a Beckmann thermometer is used at partial immersion, so the emergent-stem temperature must be monitored and any necessary correction applied. Applying the correction may be a little complicated because the lower capillary may have a different bore size than the upper one, so the same length of exposed capillary would contain a different volume of mercury. The junction of the two capillaries also presents a problem, as its volume is difficult to quantify. It is better to use the thermometer with only the upper capillary exposed. If this is not possible then the degree equivalent of the volume of mercury in the exposed part of the lower capillary and in the junction must be determined and the emergent-stem temperatures of all three sections monitored separately. The calibration report will usually quote the number of degrees emergent in each section.

The thermometer cannot necessarily be 'set' and used immediately because of the effects of temporary depression (section 5.3.2). If the setting of the thermometer is below ambient then the thermometer should be kept for a few hours at or below the setting temperature to ensure complete contraction of the bulb.

As in the bomb-calorimeter thermometer, stiction may also be a problem and the thermometer may have to be dumped or tapped before a reading is taken. However, as the thermometer is vacuous, there are no internal pressure effects to worry about.

A Beckmann thermometer is usually calibrated at a number of 'settings' and the corresponding corrections will vary, because the varying amounts of mercury in the bulb. If the thermometer is set at 25°C there will be less mercury in the bulb than when it is set at 20°C, and thus less mercury to expand when the thermometer is heated, say 4°C. Therefore, the calibration correction is more positive for a setting of 25°C than it would be for 20°C.

A calibration report may supply data for each of three settings or it may give only the corrections for one setting and a setting factor [35], which must be applied.

7.1.3   Deep-sea reversing thermometers

As the name implies, these thermometers are used for measuring deep ocean temperatures. While the thermometer can easily be lowered to the desired depth, it cannot easily be read there, so it must be brought back on board ship for reading. The problem then is to ensure that the thermometer maintains the indication it had at the appropriate depth rather than showing the temperature on the ship. The high pressures associated with great depths present another problem to the thermometer designer.

The deep-sea reversing thermometer, shown schematically in Figure 7.3, is actually two thermometers enclosed in a glass sheath. The main thermometer has a large mercury-filled bulb and an ancillary bulb linked to it by a capillary. The thermometer is vacuous and has a constriction above its main bulb, which causes the mercury column to break when the thermometer is inverted. The

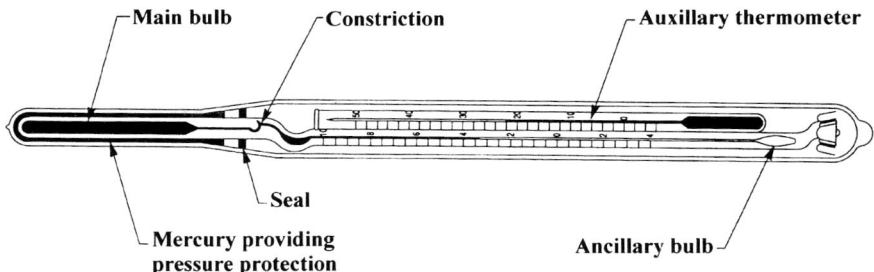

Figure 7.3: Schematic diagram of a deep-sea reversing thermometer.

auxiliary thermometer, much like a normal thermometer, has its bulb at the opposite end of the instrument from that of the main thermometer.

In use, a deep-sea reversing thermometer is put into a special cage, lowered to the required depth and allowed to equilibrate for about 20 minutes. It is then turned upside-down, by a cage mechanism, and brought back to the ocean surface. Upon inversion, the mercury separates and fills the ancillary bulb and some of the capillary. The amount separated is proportional to the temperature of the thermometer at the time of inversion—as read on the main scale. A small correction must be added to allow for the change in volume of the mercury and this is the reason for the auxiliary thermometer, which indicates the temperature of the thermometer at the time the main scale is being read.

These thermometers come in both protected and unprotected forms. Protected thermometers have, as part of the outer glass sheath, a sealed section around the bulb partially filled with mercury (see figure). This provides some protection against the effects of pressure on the reading. Unprotected thermometers are also available, and if both types are used together, a measurement of pressure as well as temperature can be obtained.

7.1.4   Maximum-indicating thermometers

A maximum-indicating thermometer is one that indicates the maximum temperature that it has reached since it was last 'set'. There are two main types of maximum thermometer. The first, shown in Figure 7.4, has an index held lightly against the wall of the capillary by a spring. When the mercury rises in the capillary it pushes the index before it, but when the thermometer cools the mercury level drops and the spring keeps the index at the highest position. The thermometer is reset using a magnet.

The second type has a constriction of the bore below the graduated portion of the stem. The constriction allows the mercury to expand freely, but is

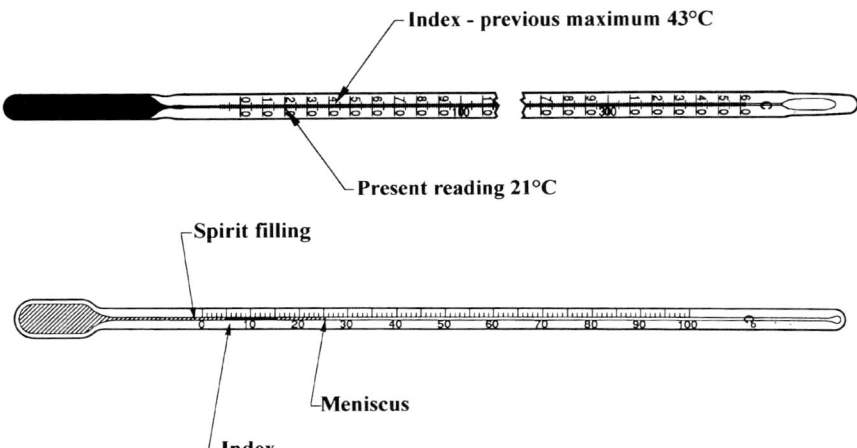

Figure 7.4: Schematics of a maximum-indicating thermometer, using mercury (upper), and a minimum-indicating thermometer (lower), which uses spirit.

sufficiently narrow to prevent it from retreating through the constriction when the thermometer cools. The top of the mercury column indicates the maximum temperature. Of course, the mercury in the separated column will also contract as the temperature decreases, so an extra correction may be required if the temperature at the time of reading is significantly different from the maximum indication.

Calibration is carried out in the usual way, but extra tests [36] are required to determine the accuracy of the maximum indication.

7.1.5   Minimum-indicating thermometers

Minimum-indicating thermometers are of the spirit-in-glass type and are used to indicate the minimum temperature reached since the thermometer was 'set' last. As they are spirit-in-glass thermometers, they have a wide bore and a precision inferior to mercury-in-glass instruments. The thermometer has a metal index, which fits loosely in the capillary below the level of the spirit, as shown in Figure 7.4. When the spirit contracts, the meniscus pulls the index down with it, but the index is left behind when the spirit expands again. It is easily reset by tilting the thermometer, causing the index to fall to the meniscus. Because the index moves freely in the spirit, the thermometer must be used in an almost-horizontal position.

Calibration is carried out in the way described for a normal spirit thermometer, but extra tests are added to determine the accuracy of the minimum indications and the ease of resetting the thermometer.

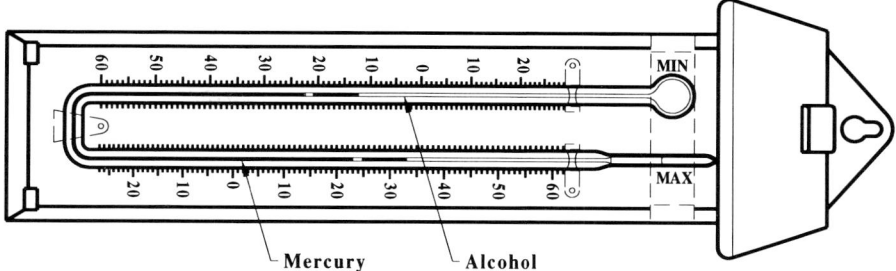

Figure 7.5: Six's thermometer.

7.1.6 Six's thermometer

In 1782, James Six developed a thermometer that indicates both the maximum and the minimum temperature reached since the thermometer was last 'set'. A modern version of Six's thermometer is illustrated in Figure 7.5. It contains a U-shaped capillary, with a bulb at each end, and with the lower part of the U filled with mercury. One bulb and arm of the thermometer (that showing the minimum value) is completely filled with spirit and the other bulb and capillary arm is partly filled with spirit. In each arm of the capillary, a metal index sits above the mercury column and is held gently in place by a spring. There are scales on both arms of the capillary: that on the partly-filled side (for the maximum value) is the right way up while that on the other side is inverted.

As the temperature increases, the mercury of a Six's thermometer is pushed towards the partly-filled side of the U, forcing the index ahead of it. When the temperature falls, the spirit in the full bulb contracts and the mercury moves up the other arm. In doing so it leaves one index behind, held in place by the spring, and pushes the second one upwards. The indices are reset using a magnet.

These thermometers are not suitable for precise measurement. One problem is that since spirit wets glass it tends to pass along the column between the mercury and the glass.

7.1.7 Clinical thermometers

The measurement of deep-body temperature is a useful aid in the diagnosis of human and animal disorders. As a consequence, a special class of maximum-indicating thermometer has been developed to measure body temperature. The performance and dimensional requirements of a clinical liquid-in-glass thermometer are given in reference [37] and a schematic diagram of one is shown in Figure 7.6.

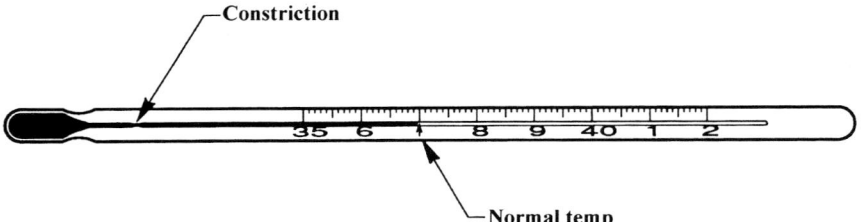

Figure 7.6: A typical liquid-in-glass clinical thermometer.

The main requirements of a clinical thermometer are that it:

- responds quickly,
- is easy to read,
- maintains its reading after removal from the body,
- is reliable and accurate,
- is able to be cleaned and disinfected and
- is safe to use in the body.

For clinical liquid-in-glass thermometers, a quick response is provided by a small bulb that rapidly comes to thermal equilibrium with the body—usually within 1 min. However, a smaller bulb has proportionally less expansion, so a finer capillary is needed to give the required accuracy and still have a scale with a sufficiently small graduation interval and easy to read. Unfortunately, a fine bore makes the mercury column difficult to see and may also lead to problems with stiction. The mercury is made easier to see by incorporating a lens front in the thermometer, which magnifies the bore when the thermometer is viewed from the appropriate angle.

These thermometers use a constriction of the bore to break the mercury column as the thermometer cools on removing it from the body. The reading of the mercury above the constriction is the measured body temperature. After the reading has been taken, the thermometer is reset by shaking the mercury down past the constriction. Obviously, correct functioning at the constriction is critical as it must cleanly break the mercury column on cooling, but must also allow the mercury to be shaken down easily. The difference between the indicated temperature and the temperature of the mercury at the time the thermometer is read is not usually large enough to require correction.

It should also be noted that clinical thermometers cannot be used where ambient temperature is higher than normal body temperature.

Clinical thermometers made to a recognised specification and used correctly are usually both reliable and accurate. Most clinical-thermometer specifications call for an uncertainty of 0.1°C or better. If necessary, the thermometer can be calibrated by comparison with a standard, as described in Chapter 6, with some additional measurements to determine its ease of resetting and the accuracy of the retained reading.

Clinical thermometers may be constructed to allow sterilisation by heating, but if the accuracy of the thermometer is to be maintained over long periods, desiccation or chemical sterilisation is usually used. The thermometer should therefore have no cavities or corners that might harbour moisture or organic material. There is obviously a further safety hazard in having poisonous material (mercury) in fragile glass placed in the body. The small bulb size and short stem reduces the hazard, but cannot entirely eliminate it.

7.2   Other liquid-expansion thermometers

Some liquid-expansion thermometers have been designed for remote reading. In these, the most commonly used filling liquid is mercury, but alcohol and xylene are also used. The thermometer comprises a bulb, which contains the liquid and is connected by a capillary to a pressure-sensitive 'Bourdon' tube. Expansion of the liquid deflects the Bourdon tube and this in turn, through a series of linkages, moves a pointer on a gauge or a pen on a recorder.

The vapour pressure of the filling liquid must be negligible over the temperature range of its use, and its coefficient of expansion must be large.

Such thermometers are not very accurate, but the facility for remote reading makes them very useful for many industrial measurements.

Chapter 8

Calibration Enclosures

Corinna Horrigan

8.1   Introduction

All types of thermometer measure temperature by measuring the temperature-related change in a specific physical property, such as volume or resistance. Not only do thermometers of different types need to have their measurements related, but measurements by individual thermometers of the same kind may differ due to variations in construction and materials and from use. Consequently, all thermometers should be calibrated, either directly at the fixed points or by comparison with a known instrument, before they can be used with confidence.

Primary standard instruments are calibrated directly on ITS–90, using the defined fixed points and the appropriate interpolation equations, and their calibrations must then be transferred to other thermometers, many of which cannot be calibrated at the fixed points. This may be because there is no interpolation function (e.g., liquid-in-glass thermometers) or because the range of the thermometer does not include enough fixed points (e.g., thermistors). For these reasons, most working thermometers are calibrated by comparison with another, previously calibrated thermometer, for which there should be a chain of calibration, traceable to ITS–90 [38].

Two types of temperature-controlled enclosure are used in the calibration of thermometers. One is the fixed-point enclosure, which is used mainly by standards laboratories for calibration at defined fixed points, while the second is the variable-temperature enclosure used by many different laboratories to calibrate thermometers by the comparison method.

The fixed-point enclosure usually has a small working space, as effective thermal transfer is required and heat losses must be minimised. Fortunately, it has to accommodate only one temperature sensor at a time. The comparison method requires a larger working space, to accommodate both a standard and at least one test thermometer at the same time. The uniformity of the enclosure will be a limitation on the calibration accuracy. Both methods require controlled heating and/or cooling to achieve the desired temperature and uniformity.

There may be a long chain of calibrations between the primary standard and the working thermometer, and this will inevitably lead to increased uncertainty. It is important that the magnitude of the uncertainty be known. The characteristics of the calibration enclosure will make a contribution to the uncertainty, and each enclosure should be tested to measure the extent of its contribution.

Both types of calibration enclosure are discussed in this chapter, although variable-temperature enclosures are discussed in more detail, as most working thermometers are calibrated using this type of enclosure. A wide variety of these enclosures is available, and the best choice is determined by factors such as temperature and uncertainty requirements, safety, space, convenience and cost.

8.2   Fixed-point enclosures

Certain equilibrium states and physical transitions, such as triple points and freezing points, occur at highly-reproducible temperatures. Some of these are used as primary reference points to define ITS–90 (see Chapter 9). Other fixed points, called secondary reference points, are not used to define the scale, but are useful as reference and check points.

In most cases, a fixed point consists of a sealed cell that contains the reference material and isolates it from contamination. High-purity material is required to obtain an accurate and reproducible point, so the cell wall usually includes a re-entrant well into which the thermometer is placed, so that it can achieve thermal equilibrium with the cell without contaminating the material. An outer enclosure (e.g., a furnace) surrounds the cell and provides the heating or cooling necessary to obtain and maintain the appropriate temperature. In some cases, one such outer enclosure may be used with a number of different fixed points, although it is usually simpler to have a separate one for each fixed point required, so that optimum heating and cooling settings can be maintained.

As the primary fixed points define the scale, they are used extensively by standards laboratories for work with primary standards. With the exception of the ice point, it is rarely economic for other laboratories to set up primary or secondary fixed points.

When the fixed points are correctly realised the temperature is defined. Thermometers being calibrated at a fixed point do not need to be compared to a standard thermometer at that temperature.

8.2.1 Triple points

The triple point of a material is the temperature at which its solid, liquid and vapour phases co-exist in thermal equilibrium at the vapour pressure of the material. If the starting material is very pure and the cells have been properly prepared and sealed, such a point can be reproducible to tenths of a millikelvin (mK).

A number of triple points are used to define the scale (see Table 9.1 on page 164). The most important is that of water, because it is used to define the Kelvin and it forms part of both the high- and the low-temperature scales. The triple point of water is defined as 273.16 K (0.01 °C). A schematic diagram of a water triple point assembly is shown in Figure 1.3.

A properly cared for water triple point can be maintained relatively easily for many months in an ice bath, but a cryostat is required to reach and maintain the low-temperature triple points. Many factors, such as isotope ratios and cell material, must be considered to obtain the highest accuracy possible at triple points. Radiative heat transfer can also be important. For example, both glass and ice are transparent at some wavelengths and the thermometer reading in the triple point may be affected by transmitted radiation.

Further information on triple points and their construction can be found in section 1.5.1 and references [39, 40].

8.2.2 Freezing points

Many of the primary and secondary fixed points of ITS–90 are based on the freezing points of materials. The freezing point is the temperature at which a substance undergoes a phase change from a liquid to a solid. This transition results in the emission of latent heat and this compensates for the heat loss until the freezing process is complete. Depending on the rate of heat loss, the point can be maintained for many hours. A typical freezing point cell is shown in Figure 1.4.

The freezing points of tin, zinc and silver are some of the primary fixed points on the scale (Table 9.1), while those of antimony and cadmium are secondary points. The ice point, which is actually a melting point, is an important reference temperature for many secondary standards and working thermometers and is discussed more fully in the next section.

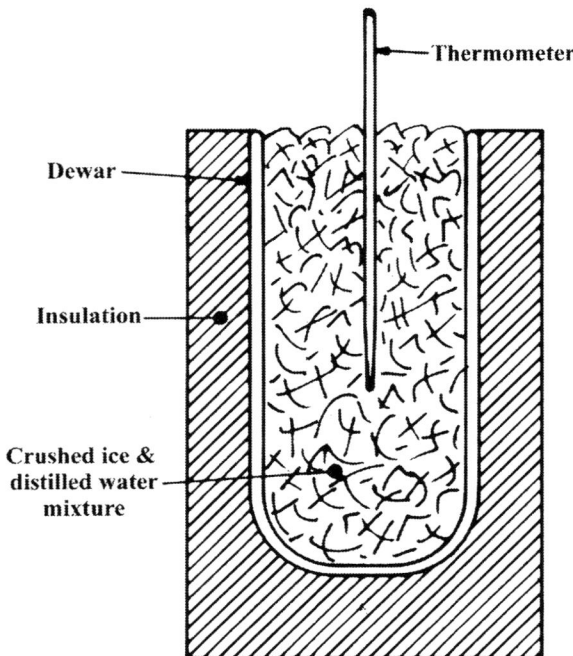

Figure 8.1: Ice-point apparatus.

As with the triple points, the freezing points can be affected by many factors, such as purity and supercooling [39, 40]. For more information see section 1.5.1).

8.2.3   The ice point

The ice point is the melting point of ice at normal atmospheric pressure. The temperature at this point is 273.15 K (0.00°C) and it is reproducible to better than 1 mK, though some care is still necessary with both material and equipment to achieve the highest accuracy. It is now a secondary reference point, having been replaced as the fundamental point by the more reproducible water triple point. Unlike most fixed points it does not require a sealed container and is fairly easy to prepare and maintain in a laboratory. This makes it ideal for use as a reference point for liquid-in-glass thermometers, platinum-resistance thermometers and other sensors whose range includes 0°C. For this reason it is probably the most widely used reference point of all.

Figure 8.1 is a schematic diagram of an ice point. To make an ice point, a well insulated, wide-mouthed container, preferably a dewar (vacuum flask), is required. It should have a stable support stand and be deep enough to permit sufficient immersion of any thermometer likely to be calibrated. The container is then filled with an ice slurry made from distilled-water ice, which has been crushed and then thoroughly mixed with just enough distilled water for the ice to lose its opacity. To reduce the risk of contamination, the container and the stirrer used in the mixing should not be used for other purposes and should be made of some insoluble material, such as glass or stainless steel. Contamination will usually lower the temperature of an ice point.

The water is held in the ice by capillarity and it helps to provide good thermal contact with the thermometer and to quickly bring the ice to equilibrium. To ensure that sufficient water is retained and that all the ice is at a uniform temperature, very small ice pieces a few mm in diameter are best. Small pieces can be obtained by shaving or by crushing thoroughly.

For some purposes, the highest accuracy is not required. Then, it is possible to use commercial ice or town water to make up an ice point, and the accuracy achieved depends on the quality of the tap water. For example, ice from Sydney tap water will not usually depress the ice-point temperature by more than 10 mK. The water supply can vary from place to place and often from season to season and it should be tested before it is used, either by checking a test ice point against known thermometers or by measuring the conductance of the water, which should be less than $100\,\mu S$. In addition, the freezing process tends to concentrate the impurities in the section that was frozen last. This leaves the first frozen ice relatively pure and it appears clear while the part containing the impurities is cloudy. If possible, only the clear ice should be used.

Depending on the container, a properly made and maintained ice point can maintain its temperature for several hours with a uniformity of better than 10 mK. It is often used for its convenience as a uniform-temperature enclosure, as well as for its fixed temperature.

However, to maintain this uniformity it does require some attention over time. The conditions in the ice pot change as the ice gradually melts. Ice melts away most quickly from the sides of the container and from the sides of any thermometers immersed in the ice. The melt water will not all be retained in the slurry and, as water is more dense than ice, it will form a pool at the bottom of the ice pot, and this can be more than 50 mK warmer than the slurry.

These regions should be avoided when making measurements. The ice pot can be refurbished by draining the excess water from the bottom of the pot using a siphon or a drain, if provided, and adding extra slurry. The slurry is then well stirred to eliminate any spaces left by other thermometers and the meltbacks at the side.

As with the triple point, transmitted radiation can significantly affect the temperature of high-accuracy thermometers, especially glass-sheathed platinum-resistance thermometers (PRT's). If necessary, the effect can be reduced by covering the top of the ice slurry with a piece of clean aluminium foil.

8.3   Variable-temperature enclosures

In one form or another, variable-temperature enclosures can be used to cover much of ITS–90. Cryostats are used for the very low temperatures up to about 0 °C, various types of stirred-liquid bath can be used from −50 °C to about 600 °C and furnaces can be used to more than 1800 °C. The range from −50 °C to 600 °C covers the most commonly used thermometers and this discussion will concentrate mainly on enclosures for this range.

8.3.1   Stirred-liquid baths

The stirred-liquid bath is a widely used calibration enclosure. Depending on the liquid used, its working range lies between −50 °C and 600 °C. For bath temperatures around ambient, the temperature variation in the working space of a well-designed bath can be as little as 1 or 2 mK and variations less than 50 mK are fairly easy to achieve.

In most calibration baths the working space is separated from the region where the heating and/or cooling take place. The stirring thoroughly mixes the liquid and circulates it between the two regions. Circulation without thorough mixing may lead to laminar flow in which hot and cold volumes of liquid can circulate separately.

There are two commonly-used designs for calibration baths. In the concentric type (Figure 8.2), the working space is isolated from the main container by an inner cylinder. The stirring pushes the liquid up the inner cylinder (working space) and it overflows into the main container to be recirculated.

The second type has two side-by-side, parallel cylinders connected at the bottom and the top. One of the cylinders represents the working space while the other contains all the heating, cooling and stirring apparatus. Circulation in this case may be in either direction. Both types of bath need top covers to reduce heat loss at the surface and, for most uses, to provide some kind of support for the thermometers.

The essential requirement of these baths is to provide a working space with a uniform temperature, and they should be thoroughly tested before use (see section 8.4). Heat losses are reduced as much as possible by appropriate

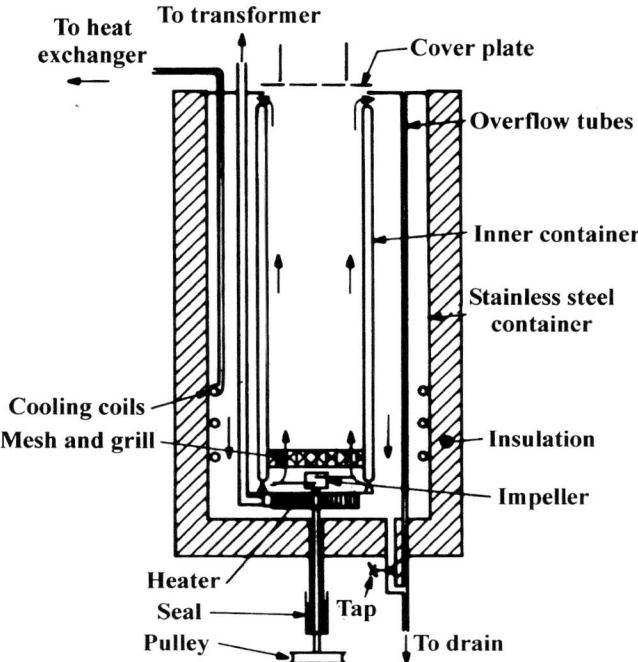

Figure 8.2: Concentric-design stirred-water bath showing the direction of flow.

insulation and are balanced by input from the heaters. The temperature uni-formity in the working area is affected by the circulation rate and the mixing. Improving these may increase the uniformity of a bath, however, it is not the answer to poor design, as there are limitations on the amount stirring can be increased without causing cavitation or overheating. Another way of reducing small temperature variations or instabilities is to use a temperature-smoothing block—a solid metal block, usually of copper or aluminium, containing wells for the thermometers. It will smooth out temperature fluctuations, but its temperature will lag behind that of the bath and this, in turn, will affect temperature control. The problem may be reduced by placing the control thermometer in the fluid surrounding the block, rather than in the block itself. However the control is set up, it is important to ensure that the block and liquid are in thermal equilibrium before any calibration is started. The time taken to reach equilibrium and the response time of the system should be determined during the initial testing of the enclosure (see section 8.4).

It is obvious that any bath to be used below room temperature will need some form of cooling, but it should be noted that cooling may also be required for bath temperatures above ambient, because of the heat input from stirring. It would not be unusual to require some cooling to maintain a bath temperature that is 20°C above ambient.

The temperature of the calibration bath must not only be uniform but, for most purposes, it should also be stable. Stability is usually achieved using an electronic controller on which the desired temperature can be set, while noting that the placement of the control sensor can affect the stability of the bath [41]. If it is too close to the heater, then the bath may be slow to reach its set temperature, while if it is too far away, the time delay may cause the temperature to oscillate.

As already mentioned, stirred-liquid baths can be used over a wide range of temperatures. Unfortunately, no single liquid will cover the entire range and some liquids may be costly or present safety problems—for a list of common bath liquids see Table 8.1.

Table 8.1: Temperature range over which common bath liquids can be used and some specific problems in their use.

Bath Liquid	Temperature Range (°C)	Problems
Pentane	−200 to −50	flammable/evaporation
Alcohol	−70 to +20	flammable/evaporation
Water/glycol	−30 to +100	
Silicone oil	+50 to +250	cost
Organic oils	+200 to +300	dangerous fumes/flammable
Salt	+260 to +600	corrosive

Stirred-liquid baths may be obtained commercially or constructed by the laboratory. The choice of design and supplier will depend on the temperature range and uniformity required as well as the type of thermometers to be calibrated.

8.3.2   Calibration furnaces

Laboratory calibration furnaces are used mainly for thermocouples and, with suitable furnace liners, can also be used for the calibration of some platinum-resistance thermometers and high-temperature liquid-in-glass thermometers. A calibration furnace is usually tubular, with the heat input distributed as evenly as possible along its length (Figure 8.3). Appropriate insulation should be used to reduce heat losses and, for high-temperature furnaces, a water-cooled jacket may be required for safety.

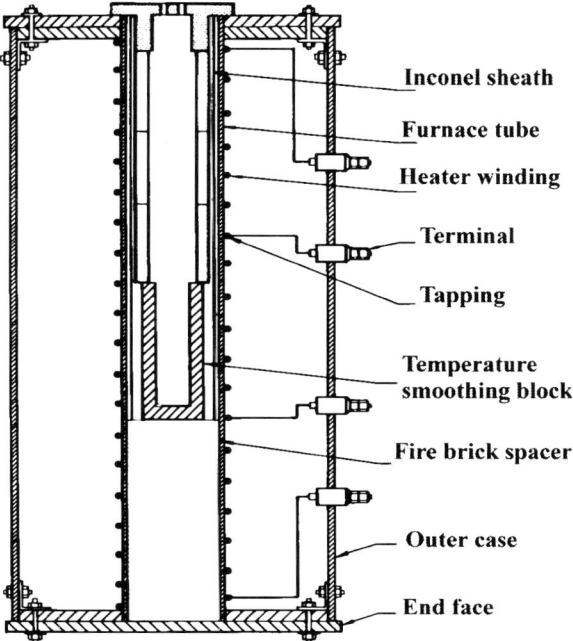

Inconel sheath

Furnace tube

Heater winding

Terminal

Tapping

Temperature smoothing block

Fire brick spacer

Outer case

End face

Figure 8.3: A calibration furnace designed for use with a metal smoothing block, showing the tapped heating element normally required.

The furnace design illustrated in Figure 8.3 is a commonly used one suitable for temperatures up to about 1200 °C, depending on the type of heater winding used [42]. Furnaces for use at temperatures higher than this have more complex heater requirements and, for very high temperatures (>1600 °C) an inert atmosphere may be required [39]. Furnace controllers are readily available commercially, although it is best to use one that provides for current limiting or some form of ramping, as heating a cold furnace too quickly can cause structural problems.

The type of furnace recommended for the calibration of thermocouples is that developed at NML [43] and discussed in reference [26]. It does not depend on a smoothing block, and thus the need to search the block for temperature uniformity, but instead relies on the tips of the standard and test thermocouples being wire-wrapped or welded together. It requires a temperature uniformity of only about 1% in the vicinity of the tips. If, however, the tips of the thermocouples are not joined or if other sensors are being calibrated, then a much better uniformity is required over a longer space, and is usually achieved by using furnace liners.

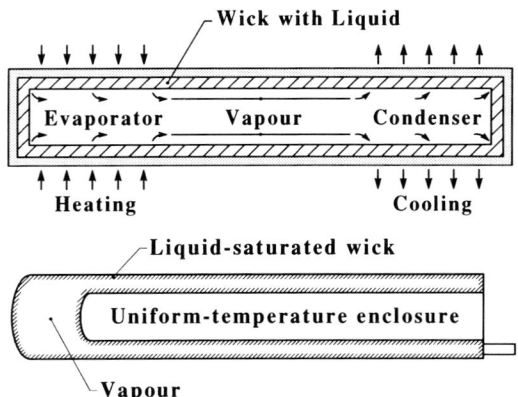

Figure 8.4: A simple heat pipe (upper) and a heat-pipe liner (lower) used as a temperature enclosure for thermometer calibrations.

For a long time, metal blocks were the main type of furnace liner available, although heat pipes are now increasingly used in this capacity. With the metal block, the high conductivity of the metal smooths out temperature variations along its length and its thermal mass smooths out small fluctuations with time. The process is not instantaneous and the relatively high thermal capacity of a metal block leads to a sluggish recovery from a cold charge, such as having thermometers inserted, and to a slow response to a change in the set temperature.

8.3.3 Heat pipes

A heat pipe transports heat by mass transfer, driven by a pressure gradient, and the effective thermal conductivity can be hundreds of times greater than that of copper. A simple heat pipe like that shown in Figure 8.4 is a sealed tube containing a small amount of volatile liquid. A small temperature rise at one end of the tube causes the liquid there to evaporate. This results in a pressure wave that propagates along the tube at close to the speed of sound. At the cooler end the vapour will condense and release its latent heat. The liquid is transported back to the hot end either by gravity or by capillary action along a wick or fine grooves cut in the inner surface. Not only does such a device transfer heat very quickly, but if a layer of liquid can be maintained in the internal surface, then temperature differences along the tube will be minimised.

To be used as a surface liner, this simple design must be modified by the addition of a central cavity that provides the working space for calibrations, as in Figure 8.4. If a film of liquid can be maintained on all the internal surfaces,

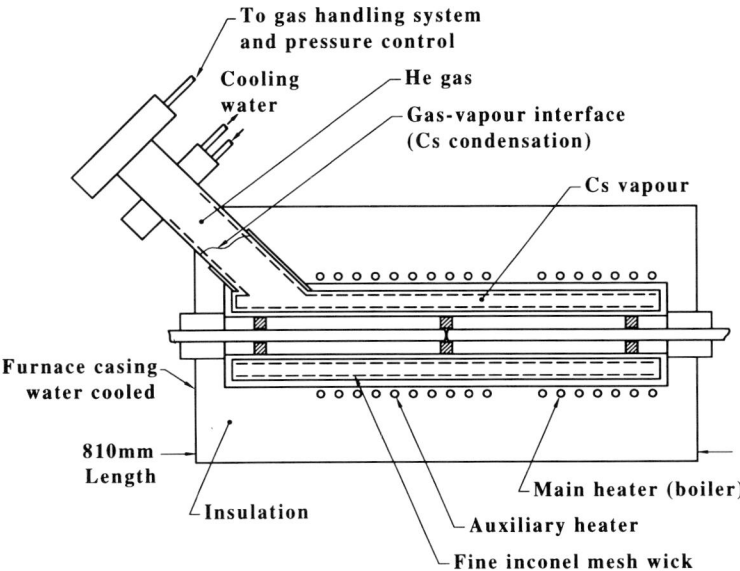

Figure 8.5: A schematic outline of a Cs heat-pipe furnace.

then a very uniform temperature can be maintained along the length of the cavity. Suitable wicking can aid this.

Such liners have been used in fixed-point, black-body and inter-comparison furnaces. It is possible to obtain a uniformity better than $\pm 5\,\mathrm{mK\,cm^{-1}}$. Temperature stability, however, is dependent of the quality of the furnace control, and radial temperature gradients may also be significant depending on the size and design of the cavity.

For any liquid, there is a unique boiling temperature corresponding to a particular pressure, and this relationship may be used for the temperature control of heat pipes. In the gas-pressure-controlled heat pipe an inert gas, usually helium, is used to control the internal pressure of the heat pipe. The lower-temperature limit is set by the availability of sufficient vapour to transport heat and the upper limit is set by the strength of the container at high temperature and pressure. A schematic of such a heat pipe is shown in Figure 8.5 and more information can be found in references [27, 44]

8.3.4 Other calibration enclosures

Calibrations below 0°C require a cryostat, which comes in a number of forms, depending on the thermometers with which it will be used. For example,

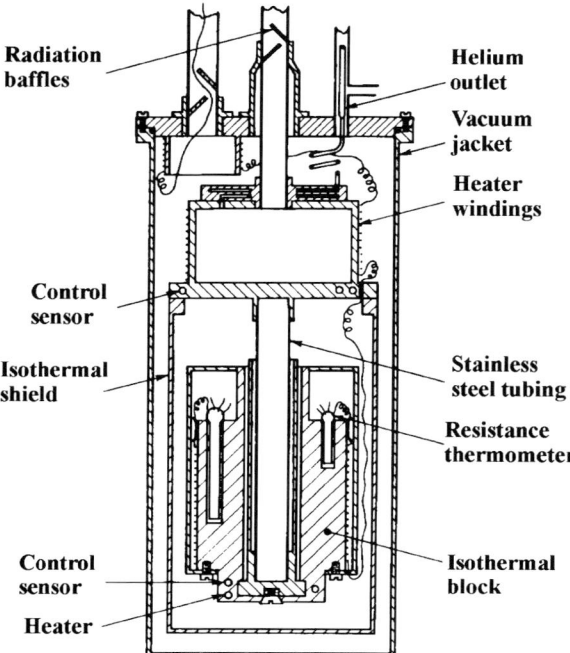

Figure 8.6: A cryostat designed for the calibration of low-temperature resistance thermometers.

the cryostat used at NML for the calibration of liquid-in-glass thermometers is basically a copper block with a number of thermometer wells containing an iso-pentane mixture and cooled by a liquid-nitrogen heat exchanger. To minimise heat transfer to its surroundings, the block is placed in a vacuum chamber and the top is insulated by polystyrene. For other forms of sensor, other designs are more appropriate. Figure 8.6 shows a cryostat designed for the calibration of low-temperature resistance thermometers. These cryostats are not in common use and more details can be found in the references [39, 45] if required.

The fluidised-bed bath is a stirred-powder bath and can be used from room temperature to about 900°C, depending on the powder used. The bath is filled with a fine powder, such as aluminium oxide, which is fluidised by forcing fine bubbles of gas through it. The enclosure can be quite dusty and, even with a temperature smoothing block, their uniformity is less than that achievable in stirred-liquid baths.

8.4 Enclosure testing

As already mentioned, an enclosure must be thoroughly tested before it can be used for calibration. The testing determines whether temperature within the working space has the required level of uniformity and whether temperature variations over time are also within acceptable limits. If the working space is large enough, a test procedure using two or more probes is best but, as this is not always possible, a single probe method may also be used. Before testing is started, the uniformity requirements and the temperature range of the enclosure should be decided. The general question of enclosure testing is covered in reference [46].

Two-probe method

Variations within an enclosure may be measured with two temperature probes of similar precision and response time. One of the sensors is held in the same position throughout the test, as a reference, while the other begins as close as possible to this position and is then moved to a number of other sites in a systematic search pattern that covers the entire working space of the enclosure (Figure 8.7). The movable probe should be left in each position long enough to reach equilibrium.

It is preferable to use thermometers with an electrical output since this can be recorded directly on a chart recorder or computer. It is the *variation in the difference* between the two outputs that is significant. However, it is important that the response times of each detector be approximately the same. An example of a chart recorder output is shown in Figure 8.8. After a search has been completed at one temperature it should be repeated at a number of other temperatures covering the whole range of use for the enclosure. For example, if a water bath is to be used to cover the range 5 to 60 °C, then possible test temperatures would be 5, 20, 40 and 60 °C, though exact temperatures do not matter for this exercise. It is more important to choose temperatures where problems are likely to occur. For example, when the amount of heating is minimal and there may be problems maintaining control.

Few baths will have temperature variations as small as those shown in Figure 8.8, which realates to a specially designed bath. However, it is necessary that the temperature variation over the working space be less than half (preferably less than one fifth) of the least uncertainty envisioned for sensors to be calibrated in that enclosure. Of course, if sensors are to be compared while close together, then the working space to be tested may be quite small and then these constraints are relatively easy to achieve.

The enclosure should also be tested for both long- and short-term temporal

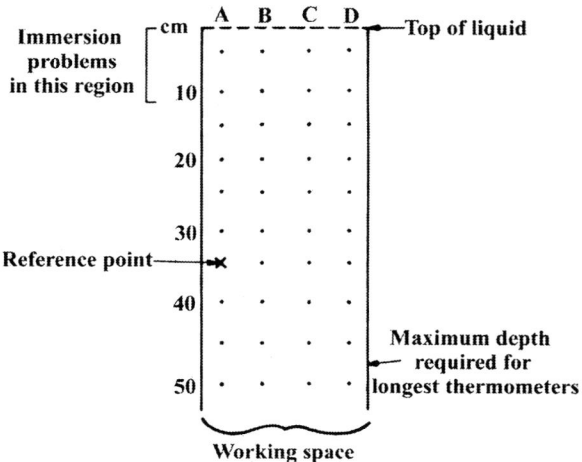

Figure 8.7: A two-dimensional schematic of a geommetric search grid within the working space of a bath, the test rows being identified by A, B, etc.

variations in temperature. This can be done by monitoring the temperature at a representative calibration site over a period of time that is at least as long as a typical calibration run at that temperature. This should reveal any oscillation in the control system and any drift that would affect a normal calibration. Once again, this should be repeated at a number of different temperatures and it may be possible to combine it with the search for spatial temperature variations. Naturally, it is important to use a thermometer that is itself stable over time at the temperatures being tested.

Long-term fluctuations or low-amplitude drift may not significantly affect the use of a calibration enclosure, depending on the type of calibration being carried out. However, the extent of such variation should be measured and its possible effects determined and included in the calculation of uncertainty. Short-term oscillation on the other hand can cause quite significant errors especially if it is on a similar time scale as the response time of the standard and/or the thermometer being calibrated.

Figure 8.9 shows such a situation. It is possible in these circumstances to have the temperature of the standard increasing while that of the test thermometer is decreasing, as at the point marked A. Obviously, if the magnitude of this oscillation is no greater than the sensitivity of the thermometer, then it will cause no problems.

One point to remember while testing stirred-liquid baths is that the mixing characteristics may be changed when something, such as a thermometer-support cage, is immersed in it. For this reason, it is best to keep the test setup as close as possible to that used for calibration.

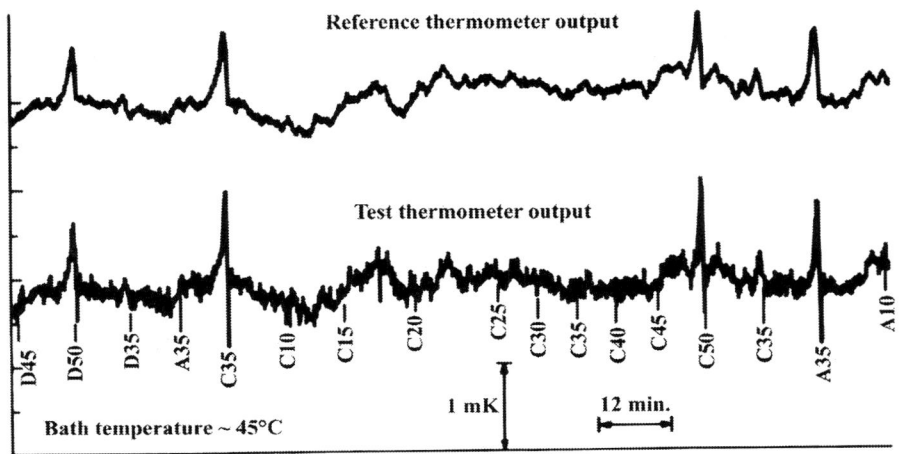

Figure 8.8: Typical chart-recorder trace obtained during the search of a water bath. The labels A35, D35 etc. refer to the positions defined in Figure 8.7.

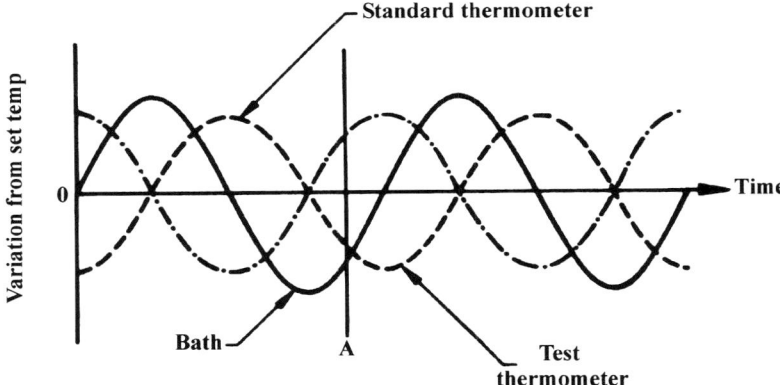

Figure 8.9: Short-term oscillations in bath temperature and the corresponding effect on the temperatures of thermometers with different response times.

In some cases, it may be possible and appropriate to use a multiple-probe set-up, such as a set of differential thermocouples. The same basic concepts apply. Other types of thermometer can sometimes be used for enclosure testing, especially if the uncertainty requirements of the enclosure are not exacting. For instance, if the uncertainty requirement is 10 mK or more, then differential thermocouples [47] may be used.

Single-probe testing

Testing an enclosure with a single temperature probe uses the same principles as two- or multiple-probe testing, but requires more patience. As with the other method, a particular position is designated as a reference site. The probe is first placed at this site and allowed to reach equilibrium. It is then moved to a second search site until equilibrium is reached again. Then it must be returned to the reference position. This is repeated for each test site in the enclosure. Data for the reference site will show any systematic changes over time (temporal variation). The spatial variation is determined from the mean temperatures obtained at each test site, corrected for drift in time using data at the reference site. Once again, the test should be repeated for a number of different temperatures covering the range of use for the enclosure.

8.5   Setting up a calibration facility

To avoid the purchase of unsuitable equipment, calibration needs should be realistically evaluated and the following points considered.

(a) Temperature range to be covered. Will more than one enclosure be required? What kinds of enclosure are needed?

(b) Uncertainty requirements in each temperature range. It is important to be realistic here! Don't try for a higher accuracy than you require.

(c) Depth required for correct immersion of thermometers. PRT's require significant immersion (at least 150 mm) and total-immersion, liquid-in-glass thermometers can be up to 400 mm long.

(d) Special control requirements. Some calibration methods require slowly increasing temperatures at a test point and some equipment needs special control to avoid damage.

(e) Laboratory environment. Where will the enclosure be situated? Will it require air conditioning (temperature and humidity?), a fume hood or other safety measures? Will it require cooling water or drainage?

(f) Cost. How much can you afford? Remember that it is a false economy to buy something that is cheap if it does not meet your needs.

(g) Maintenance and service. Are there local service facilities? Can you maintain it yourself?

Chapter 9

International Temperature Scale of 1990

A version of the official text [48]

The International Temperature Scale of 1990 was adopted by the International Committee of Weights and Measures at its meeting in 1989, in accordance with the request embodied in Resolution 7 of the 18th General Conference of Weights and Measures of 1987. This scale supersedes the International Practical Temperature Scale of 1968 (amended edition of 1975) and the 1976 Provisional 0.5 K to 30 K Temperature Scale.

9.1 Units of temperature

The unit of the fundamental physical quantity known as thermodynamic temperature, symbol T, is the kelvin, symbol K, defined as the fraction $1/273.16$ of the thermodynamic temperature of the triple point of water[1].

Because of the way earlier temperature scales were defined, it remains common practice to express a temperature in terms of its difference from 273.15 K, the ice point. A thermodynamic temperature, T, expressed in this way is known as a Celsius temperature, symbol t, defined by:

$$t/°C = T/K - 273.15 \tag{9.1}$$

---

[1]Comptes Rendus des Séances de la Treizième Conférence Générale des Poids et Mesures (1967-1968), Resolutions 3 and 4, p.104.

The unit of Celsius temperature is the degree Celsius, symbol °C, which is by definition equal in magnitude to the kelvin. A difference of temperature may be expressed in kelvins or degrees Celsius.

The International Temperature Scale of 1990 (ITS–90) defines both International Kelvin Temperatures, symbol T_{90}, and International Celsius Temperatures, t_{90}. The relation between T_{90} and t_{90} is the same as that between T and t, i.e. :

$$t_{90}/°C = T_{90}/K - 273.15 \qquad (9.2)$$

The unit of the physical quantity T_{90} is the kelvin, symbol K, and the unit of the physical quantity t_{90} is the degree Celsius, symbol °C, as is the case for the thermodynamic temperature T and the Celsius temperature t.

9.2   Principles of the International Temperature Scale of 1990 (ITS–90)

The ITS–90 extends upwards from 0.65 K to the highest temperature practicably measurable in terms of the Planck radiation law using monochromatic radiation. The ITS–90 comprises a number of ranges and sub-ranges throughout each of which temperatures T_{90} are defined. Several of these ranges or sub-ranges overlap, and where such overlapping occurs differing definitions of T_{90} exist: these differing definitions have equal status. For measurements of the very highest precision there may be detectable numerical differences between measurements made at the same temperature but in accordance with differing definitions. Similarly, even using one definition, at a temperature between defining fixed points two acceptable interpolating instruments (e.g. resistance thermometers) may give detectably differing numerical values of T_{90}. In virtually all cases these differences are of negligible practical importance and are at the minimum level consistent with a scale of no more than reasonable complexity: for further information on this point, see reference [20].

The ITS–90 has been constructed in such a way that, throughout its range, for any given temperature the numerical value of T_{90} is a close approximation to the numerical value of T according to best estimates at the time the scale was adopted. By comparison with direct measurements of thermodynamic temperatures, measurements of T_{90} are more easily made, are more precise and are highly reproducible.

There are significant numerical differences between the values of T_{90} and the corresponding values of T_{68} measured on the International Practical Temperature Scale of 1968 (IPTS–68), see Table 9.6. Similarly, there were differences between the IPTS–68 and the International Practical Temperature

Scale of 1948 (IPTS–48), and between the International Temperature Scale of 1948 (ITS–48) and the International Temperature Scale of 1927 (ITS–27). See section 9.4 and reference [20] for more detailed information.

9.3 Definition of the International Temperature Scale of 1990

Between 0.65 K and 5.0 K T_{90} is defined in terms of the vapour-pressure temperature relations of ^{3}He and ^{4}He.

Between 3.0 K and the triple point of neon (24.5561 K) T_{90} is defined by means of a helium gas thermometer calibrated at three experimentally realisable temperatures having assigned numerical values (defining fixed points) and using specified interpolation procedures.

Between the triple point of equilibrium hydrogen (13.8033 K) and the freezing point of silver (961.78°C) T_{90} is defined by means of platinum resistance thermometers calibrated at specified sets of defining fixed points and using specified interpolation procedures.

Above the freezing point of silver (961.78°C) T_{90} is defined in terms of a defining fixed point and the Planck radiation law.

The defining fixed points of the ITS–90 are listed in Table 9.1. The effects of pressure, arising from significant depths of immersion of the sensor or from other causes, on the temperature of most of these points are given in Table 9.2.

9.3.1 From 0.65 to 5.0 K: helium vapour-pressure temperature equations

In this range T_{90} is defined in terms of the vapour pressure p of ^{3}He and ^{4}He using equations of the form:

$$T_{90}/K = A_o + \sum_{i=1}^{9} A_i[(ln(p/Pa) - B)/C]^i \qquad (9.3)$$

The values of the constants A_o, A_i, B and C are given in Table 9.3 for ^{3}He in the range 0.65 K to 3.2 K, and for ^{4}He in the ranges 1.25 K to 2.1768 K (the λ point) and 2.1768 K to 5.0 K.

9.3.2 From 3.0 K to the triple point of Neon (24.5561 K): gas thermometer

In this range T_{90} is defined in terms of a ^{3}He or a ^{4}He gas thermometer of the constant-volume type that has been calibrated at three temperatures.

Table 9.1: Defining fixed points of the ITS-90.

| No. | Temperature | | Substance[a] | State[b] | $W_r(T_{90})$ |
	T_{90} (K)	t_{90} (°C)			
1	3 to 5	−270.15 to −268.15	He	V	
2	13.8033	−259.3467	e-H$_2$	T	0.001 190 07
3	≈ 17	≈ −256.15	e-H$_2$ (or He)	V (or G)	
4	≈ 20.3	≈ −252.85	e-H$_2$ (or He)	V (or G)	
5	24.5561	−248.5939	Ne	T	0.008 449 74
6	54.3584	−218.7916	O$_2$	T	0.091 718 04
7	83.8058	−189.3442	Ar	T	0.215 859 75
8	234.3156	−38.8344	Hg	T	0.844 142 11
9	273.16	0.01	H$_2$O	T	1.000 000 00
10	302.9146	29.7646	Ga	M	1.118 138 89
11	429.7485	156.5985	In	F	1.609 801 85
12	505.078	231.928	Sn	F	1.892 797 68
13	692.677	419.527	Zn	F	2.568 917 30
14	933.473	660.323	Al	F	3.376 008 60
15	1234.93	961.78	Ag	F	4.286 420 53
16	1337.33	1064.18	Au	F	
17	1357.77	1084.62	Cu	F	

[a] All substances except ^{3}He are of natural isotopic composition: e−H$_2$ is hydrogen at the equilibrium concentration of the ortho- and para-molecular forms.

[b] For complete definitions and advice on the realisation of these various states, see reference [20]. The symbols have the following meanings:

   V : vapour pressure point

   T : triple point (temperature at which the solid, liquid and vapour phases are in equil.)

   G : gas thermometer point

   M,F : melting point, freezing point (temperature, at a pressure of 101 325 Pa, at which the solid and liquid phases are in equilibrium.

These are the triple point of neon (24.5561 K), the triple point of equilibrium hydrogen (13.8033 K), and a temperature between 3.0 K and 5.0 K. This last temperature is determined using a ^{3}He or a ^{4}He vapour pressure thermometer as specified in section 9.3.1.

From 4.2 K to the triple point of Neon (24.5561 K) with ^{4}He as the thermometric gas

In this range T_{90} is defined by the relation:

$$T_{90} = a + bp + cp^2 \tag{9.4}$$

where p is the pressure in the gas thermometer and a, b and c are coefficients, the numerical values of which are obtained from measurements made at the three defining fixed points given in section 9.3.2, but with the further restriction that the lowest one of these points lies between 4.2 K and 5.0 K.

Table 9.2: Effect of pressure on the temperature of some defining fixed points. The reference pressure for melting and freezing points is the standard atmosphere ($p_o = 101\,325\,\text{Pa}$). For triple points (T) the pressure effect is a consequence only of the hydrostatic head of liquid in the cell.

Substance	Assigned Value of Equilibrium Temperature T_{90} (K)	Pressure Effect (K Pa^{-1} × 10^{-8})^a	(mK per m of liquid)
e-Hydrogen (T)	13.8033	34.0	0.25
Neon (T)	24.5561	16.0	1.9
Oxygen (T)	54.3584	12.0	1.5
Argon (T)	83.8058	25.0	3.3
Mercury (T)	234.3156	5.4	7.1
Water (T)	273.16	−7.5	−0.73
Gallium	302.9146	−2.0	−1.2
Indium	429.7485	4.9	3.3
Tin	505.078	3.3	2.2
Zinc	692.677	4.3	2.7
Aluminium	933.473	7.0	1.6
Silver	1234.93	6.0	5.4
Gold	1337.33	6.1	10.0
Copper	1357.77	3.3	2.6

^{a} Equivalent to millikelvins per standard atmosphere.

From 3.0 K to the triple point of Neon (24.5561 K) with ^{3}He or ^{4}He as the thermometic gas

For a ^{3}He gas thermometer, and for a ^{4}He gas thermometer used below 4.2 K, the non-ideality of the gas must be accounted for explicitly, using the appropriate second virial coefficient $B_3(T_{90})$ or $B_4(T_{90})$. In this range T_{90} is defined by the relation:

$$T_{90} = \frac{a + bp + cp^2}{1 + B_x(T_{90})N/V} \qquad (9.5)$$

where p is the pressure in the gas thermometer, a, b and c are coefficients the numerical values of which are obtained from measurements at three defining temperatures as given in section 9.3.2, N/V is the gas density with N being the quantity of gas and V the volume of the bulb, x is 3 or 4 according to the isotope used, and the values of the second virial coefficients are given by the relations:

For ^{3}He,
$$B_3(T_{90})/m^3 mol^{-1} = \{16.69 - 336.98(T_{90}/\text{K})^{-1}$$
$$+ \ 91.04(T_{90}/\text{K})^{-2} - 13.82(T_{90}/\text{K})^{-3}\}10^{-6}. \qquad (9.6)$$

For ^{4}He,

$$
\begin{aligned}
B_4(T_{90})/m^3 mol^{-1} = \ &\{16.708 - 374.05(T_{90}/K)^{-1} - 383.53(T_{90}/K)^{-2} \\
&+ 1799.2(T_{90}/K)^{-3} - 4033.2(T_{90}/K)^{-4} \\
&+ 3252.8(T_{90}/K)^{-5}\}10^{-6}.
\end{aligned}
$$

Table 9.3: Values of the constants for the helium vapour-pressure equation (9.3), for each of the three temperature ranges for which it is valid.

Constant	^{3}He 0.65 to 3.2 K	^{4}He 1.25 to 2.1768 K	^{4}He 2.1768 to 5.0 K
A_0	1.053447	1.392408	3.146631
A_1	0.980106	0.527153	1.357655
A_2	0.676380	0.166756	0.413923
A_3	0.372692	0.050988	0.091159
A_4	0.151656	0.026514	0.016349
A_5	-0.002263	0.001975	0.001826
A_6	0.006596	-0.017976	-0.004325
A_7	0.088966	0.005409	-0.004973
A_8	-0.004770	0.013259	0
A_9	-0.054943	0	0
B	7.3	5.6	10.3
C	4.3	2.9	1.9

The accuracy with which T_{90} can be realised using equations (9.4) and (9.5) depends on the design of the gas thermometer and the gas density used. Design criteria and current good practice required to achieve a selected accuracy are given in reference [20].

9.3.3 The triple point of equilibrium hydrogen (13.8033 K) to the freezing point of Silver (961.78°C): platinum resistance thermometer

In this range T_{90} is defined by means of a platinum resistance thermometer calibrated at specified sets of defining fixed points, and using specified reference and deviation functions for interpolation at intervening temperatures.

No single platinum resistance thermometer can provide high accuracy, or is even likely to be usable, over all of the temperature range 13.8033 K to 961.78°C. The choice of temperature range, or ranges, from among those listed below for which a particular thermometer can be used is normally limited by its construction.

For practical details and current good practice, in particular concerning types of thermometer available, their acceptable operating ranges, probable accuracies, permissible leakage resistance, resistance values, and thermal treatment, see reference [20]. It is particularly important to take account of the appropriate heat treatments that should be followed each time a platinum resistance thermometer is subjected to a temperature above about 420°C.

Temperatures are determined in terms of the ratio of the resistance $R(T_{90})$ at a temperature T_{90} and the resistance $R(273.16\,\mathrm{K})$ at the triple point of water. This ratio, $W(T_{90})$, is [2]:

$$W(T_{90}) = R(T_{90})/R(273.16\,\mathrm{K}) \tag{9.7}$$

An acceptable platinum resistance thermometer must be made from pure, strain-free platinum, and it must satisfy at least one of the following two relations:

$$W(29.7646°\mathrm{C}) \geq 1.118\ 07 \tag{9.8}$$
$$W(-38.8344°\mathrm{C}) \leq 0.844\ 235$$

An acceptable platinum resistance thermometer that is to be used up to the freezing point of silver must also satisfy the relation:

$$W(961.78°\mathrm{C}) \geq 4.2844$$

In each of the resistance thermometer ranges, T_{90} is obtained from $W_r(T_{90})$ as given by the appropriate reference function, the inverse equivalents to equation (9.9) or (9.10), and the deviation $W(T_{90}) - W_r(T_{90})$. At the defining fixed points this deviation is obtained directly from the calibration of the thermometer: at intermediate temperatures it is obtained by means of the appropriate deviation function, equation (9.12), (9.13) or (9.14).

(i) For the range 13.8033 K to 273.16 K the following reference function is defined:

$$ln[W_r(T_{90})] = A_o + \sum_{i=1}^{12} A_i \left\{ [ln(T_{90}/273.16\,\mathrm{K}) + 1.5]/1.5 \right\}^i \tag{9.9}$$

An inverse function, equivalent to equation (9.9) to within 0.1 mK, is:

$$T_{90}/273.16\,\mathrm{K} = B_o + \sum_{i=1}^{15} B_i \left\{ \frac{W_r(T_{90})^{1/6} - 0.65}{0.35} \right\}^i$$

The values of the constants A_o, B_o, A_i and B_i are given in Table 9.4.

A thermometer may be calibrated for use throughout this range or, using progressively fewer calibration points, for ranges with low temperature

Table 9.4: The constants A_0, A_i, B_0, B_i, C_0, C_i, D_0 and D_i in the reference functions of equations (9.9) and (9.10).

Constant	Value	Constant	Value	Constant	Value
A_0	$-2.135\ 347\ 29$	B_0	$0.183\ 324\ 722$	C_0	$2.781\ 572\ 54$
A_1	$3.183\ 247\ 20$	B_1	$0.240\ 975\ 303$	C_1	$1.646\ 509\ 16$
A_2	$-1.801\ 435\ 97$	B_2	$0.209\ 108\ 771$	C_2	$-0.137\ 143\ 90$
A_3	$0.717\ 272\ 04$	B_3	$0.190\ 439\ 972$	C_3	$-0.006\ 497\ 67$
A_4	$0.503\ 440\ 27$	B_4	$0.142\ 648\ 498$	C_4	$-0.002\ 344\ 44$
A_5	$-0.618\ 993\ 95$	B_5	$0.077\ 993\ 465$	C_5	$0.005\ 118\ 68$
A_6	$-0.053\ 323\ 22$	B_6	$0.012\ 475\ 611$	C_6	$0.001\ 879\ 82$
A_7	$0.280\ 213\ 62$	B_7	$-0.032\ 267\ 127$	C_7	$-0.002\ 044\ 72$
A_8	$0.107\ 152\ 24$	B_8	$-0.075\ 291\ 522$	C_8	$-0.000\ 461\ 22$
A_9	$-0.293\ 028\ 65$	B_9	$-0.056\ 470\ 670$	C_9	$0.000\ 457\ 24$
A_{10}	$0.044\ 598\ 72$	B_{10}	$0.076\ 201\ 285$		
A_{11}	$0.118\ 686\ 32$	B_{11}	$0.123\ 893\ 204$	D_0	$439.\ 932\ 854$
A_{12}	$-0.052\ 481\ 34$	B_{12}	$-0.029\ 201\ 193$	D_1	$472.418\ 020$
		B_{13}	$-0.091\ 173\ 542$	D_2	$37.684\ 494$
		B_{14}	$0.001\ 317\ 696$	D_3	$7.472\ 018$
		B_{15}	$0.026\ 025\ 526$	D_4	$2.920\ 828$
				D_5	$0.005\ 184$
				D_6	$-0.963\ 864$
				D_7	$-0.188\ 732$
				D_8	$0.191\ 203$
				D_9	$0.049\ 025$

limits of 24.5561 K, 54.3584 K and 83.8058 K, all having an upper limit of 273.16 K.

(ii) For the range 0°C to 961.78°C the following reference function is defined:

$$W_r(T_{90}) = C_o + \sum_{i=1}^{9} C_i \left\{ \frac{T_{90}/\text{K} - 754.15}{481} \right\}^i \qquad (9.10)$$

An inverse function, equivalent to equation (9.10) to within 0.13 mK is:

$$T_{90}/\text{K} - 273.15 = D_o + \sum_{i=1}^{9} D_i \left\{ \frac{W_r(T_{90}) - 2.64}{1.64} \right\}^i$$

The values of the constants C_o, D_o, C_i and D_i are given in Table 9.4.

A thermometer may be calibrated for use throughout this range or, using fewer calibration points, for ranges with upper limits of 660.323, 419.527, 231.928, 156.5985 or 29.7646°C, all having a lower limit of 0°C.

---

[2]Note that this definition of $W(T_{90})$ differs from the corresponding definition used in the ITS–27, ITS–48, IPTS–48, and IPTS–68: for all of these earlier scales $W(T)$ was defined in terms of a reference temperature of 0°C, which since 1954 has itself been defined as 273.15 K.

(iii) A thermometer may be calibrated for use in the range 234.3156 K ($-38.8344\,°\mathrm{C}$) to $29.7646\,°\mathrm{C}$, the calibration being made at these temperatures and at the triple point of water. Both reference functions, equations (9.9) and (9.10), are required to cover this range.

The defining fixed points and deviation functions for the various ranges are given below, and in summary form in Table 9.5.

The triple point of equilibrium hydrogen (13.8033 K) to the triple point of water (273.16 K)

The thermometer is calibrated at the triple points of equilibrium hydrogen (13.8033 K), neon (24.5561 K), oxygen (54.3584 K), argon (83.8058 K), mercury (234.3156 K), and water (273.16 K), and at two additional temperatures close to 17.0 K and 20.3 K. These last two may be determined either: by using a gas thermometer as described in section 9.3.2, in which case the two temperatures must lie within the ranges 16.9 K to 17.1 K and 20.2 K to 20.4 K respectively; or by using the vapour pressure-temperature relation of equilibrium hydrogen, in which case the two temperatures must lie within the ranges 17.025 K to 17.045 K and 20.26 K to 20.28 K respectively, with the precise values being determined from the following respective equations:

$$T_{90}/\mathrm{K} - 17.035 \;=\; (p/kPa - 33.3213)/13.32 \qquad (9.11)$$
$$T_{90}/\mathrm{K} - 20.27 \;=\; (p/kPa - 101.292)/30$$

The deviation function is[3]:

$$W(T_{90}) - W_r(T_{90}) = a[W(T_{90}) - 1] + b[W(T_{90}) - 1]^2 + \sum_{i=1}^{5} c_i[lnW(T_{90})]^{i+n} \qquad (9.12)$$

with values for the coefficients a, b and c_i being obtained from measurements at the defining fixed points and with $n = 2$.

For this range and for the following sub-ranges the required values of $W_r(T_{90})$ are obtained from equation (9.9) or from Table 9.1.

The triple point of neon (24.5561 K) to the triple point of water (273.16 K): the thermometer is calibrated at the triple points of equilibrium hydrogen (13.8033 K), neon (24.5561 K), oxygen (54.3584 K), argon (83.8058 K), mercury (234.3156 K) and water (273.16 K). The deviation function is given by equation (9.12) with values for coefficients a, b, c_1, c_2 and c_3 being obtained from measurements at the defining fixed points and with $c_4, c_5, n = 0$.

---

[3]This deviation function, and those of equations (9.13) and (9.14), may be expressed in terms of W_r rather than W [20].

Table 9.5: Deviation functions and calibration points for platinum resistance thermometers in the various ranges in which they define T_{90}.

(a) Ranges with an upper limit of 273.16 K:

Page	Lower Limit	Deviation Functions	Calibration Points (see Table 9.1)
169	13.8033 K	equation (9.12)	2–9
169	24.5561 K	as above with $c_4 = c_5 = n = 0$	2, 5–9
170	54.3584 K	as above with c_2, c_3, c_4, $c_5 = 0$ and $n = 1$	6–9
170	83.8058 K	equation (9.13)	7–9

(b) Ranges with a lower limit of 0°C:

Page	Upper Limit	Deviation Functions †	Calibration Points (see Table 9.1)
171	961.78°C	equation (9.14)	9, 12–15
171	660.323°C	as above with $d = 0$	9, 12–14
171	419.527°C	as above with c, $d = 0$	9, 12, 13
171	231.928°C	as above with c, $d = 0$	9, 11, 12
172	156.5985°C	as above with b, c, $d = 0$	9, 11
172	29.7646°C	as above with b, c, $d = 0$	9, 10

(c) Range from 234.3156 K (-38.8344°C) to 29.7646°C:

Page		Deviation Functions	Calibration Points (see Table 9.1)
172		equation (9.14) with c, $d = 0$	8–10

† Calibration points 9 and 12–14 are used with $d = 0$ for $t_{90} \leq 660.323$°C; the values of a, b and c thus obtained are retained for $t_{90} > 660.323$°C, with d being determined from calibration point 15.

The triple point of oxygen (54.3584 K) to the triple point of water (273.16 K): the thermometer is calibrated at the triple points of oxygen (54.3584 K), argon (83.8058 K), mercury (234.3156 K) and water (273.16 K). The deviation function is given by equation (9.12) with values for the coefficients a, b and c_1 being obtained from measurements at the defining fixed points, with c_2, c_3, c_4, $c_5 = 0$ and with $n = 1$.

The triple point of argon (83.8058 K) to the triple point of water (273.16 K): the thermometer is calibrated at the triple points of argon (83.8058 K), mercury (234.3156 K) and water (273.16 K). The deviation function is:

$$W(T_{90}) - W_r(T_{90}) = a[W(T_{90}) - 1] + b[W(T_{90}) - 1]lnW(T_{90})$$

$$(9.13)$$

with the values of a and b being obtained from measurements at the defining fixed points.

From 0°C to the freezing point of silver (961.78°C)

The thermometer is calibrated at the triple point of water (0.01°C), and at the freezing points of tin (231.928°C), zinc (419.527°C), aluminium (660.323°C) and silver (961.78°C).

The deviation function is:

$$\begin{aligned} W(T_{90}) - W_r(T_{90}) & = & a[W(T_{90}) - 1] + b[W(T_{90}) - 1]^2 & \qquad (9.14) \\ & + & c[W(T_{90}) - 1]^3 + d[W(T_{90}) - W(660.323°C)]^2 \end{aligned}$$

For temperatures below the freezing point of aluminium $d = 0$, with the values of a, b and c being determined from the measured deviations from $W_r(T_{90})$ at the freezing point of tin, zinc and aluminium. From the freezing point of aluminium to the freezing point of silver the above values of a, b and c are retained and the value of d is determined from the measured deviation from $W_r(T_{90})$ at the freezing point of silver.

For this range and for the following sub-ranges the required values for $W_r(T_{90})$ are obtained from equation (9.10) or from Table 9.1.

From 0°C to the freezing point of aluminium (660.323°C): the thermometer is calibrated at the triple point of water (0.01°C), and at the freezing points of tin (231.928°C), zinc (419.527°C) and aluminium (660.323°C). The deviation function is given by equation (9.14), with the values of a, b and c being determined from measurements at the defining fixed points and with $d = 0$.

From 0°C to the freezing point of zinc (419.527°C): the thermometer is calibrated at the triple point of water (0.01°C), and at the freezing points of tin (231.928°C) and zinc (419.527°C). The deviation function is given by equation (9.14) with the values of a and b being obtained from measurements at the defining fixed points and with $c, d = 0$.

From 0°C to the freezing point of tin (231.928 °C: the thermometer is calibrated at the triple point of water (0.01°C), and at the freezing points of indium (156.5985°C) and tin (231.928°C). The deviation function is given by equation (9.14) with the values of a and b being obtained from the measurements at the defining fixed points and with $c, d = 0$.

From 0°C to the freezing point of indium (156.5985°C): the thermometer is calibrated at the triple point of water (0.01°C), and at the freezing point of indium (156.5985°C). The deviation function is given by equation (9.14) with the value of a being obtained from measurements at the defining fixed points and with b, c, $d = 0$.

From 0°C to the melting point of gallium (29.7646°C): the thermometer is calibrated at the triple point of water (0.01°C), and at the melting point of gallium (29.7646°C). The deviation function is given by equation (9.14) with the value of a being obtained from measurements at the defining fixed points and with b, c, $d = 0$.

The triple point of mercury (-38.8344°C) to the melting point of gallium (29.7646°C)

The thermometer is calibrated at the triple points of mercury (-38.8344°C), and water (0.01°C), and at the melting point of gallium (29.7646°C).

The deviation function is given by equation (9.14) with the values of a and b being obtained from measurements at the defining fixed points and with c, $d = 0$.

The required values of $W_r(T_{90})$ are obtained from equations (9.9) and (9.10) for measurements below and above 273.16 K respectively, or from Table 9.1.

9.3.4   The range above the freezing point of silver (961.78°C): Planck's radiation law

Above the freezing point of silver the temperature T_{90} is defined by the equation:

$$\frac{L_\lambda(T_{90})}{L_\lambda[T_{90}(X)]} = \frac{exp(c_2[\lambda T_{90}(X)]^{-1}) - 1}{exp(c_2[\lambda T_{90}]^{-1}) - 1} \tag{9.15}$$

where $T_{90}(X)$ refers to any one of the silver $\{T_{90}(Ag) = 1234.93\,\text{K}\}$, the gold $\{T_{90}(Au) = 1337.33\,\text{K}\}$ or the copper $\{T_{90}(Cu) = 1357.77\,\text{K}\}$ freezing points [4] and in which $L_\lambda(T_{90})$ and $L_\lambda[T_{90}(X)]$ are the spectral concentrations of the radiance of a blackbody (*in vacuo*) at the wavelength λ, at T_{90} and at $T_{90}(X)$ respectively, and $c_2 = 0.014388\,\text{m·K}$.

For practical details and current good practice for optical pyrometry, see reference [20].

---

[4]The T_{90} values of the freezing points of silver, gold and copper are believed to be self consistent to such a degree that the substitution of any one of them in place of one of the other two as the reference temperature $T_{90}(X)$ will not result in significant differences in the measured values of T_{90}.

9.4 History of international temperature scales

9.4.1 International Temperature Scale of 1927 (ITS–27)

The International Temperature Scale of 1927 was adopted by the seventh General Conference of Weights and Measures to overcome the practical difficulties of the direct realisation of thermodynamic temperatures by gas thermometry, and as a universally acceptable replacement for the differing existing national temperature scales. The ITS–27 was formulated so as to allow measurements of temperature to be made precisely and reproducibly, with as close an approximation to thermodynamic temperatures as could be determined at that time. Between the oxygen boiling point and the gold freezing point it was based upon a number of reproducible temperatures, or fixed points, to which numerical values were assigned, and two standard interpolating instruments.

Each of these interpolating instruments was calibrated at several of the fixed points, this giving the constants for the interpolating formula in the appropriate temperature range. A platinum resistance thermometer was used for the lower part and a platinum rhodium/platinum thermocouple for temperatures above 660°C. For the region above the gold freezing point, temperatures were defined in terms of the Wien radiation law: in practice, this invariably resulted in the selection of an optical pyrometer as the realising instrument.

9.4.2 International Temperature Scale of 1948 (ITS–48)

The International Temperature Scale of 1948 was adopted by the ninth General Conference. Changes from the ITS–27 were: the lower limit of the platinum resistance thermometer range was changed from −190°C to the defined oxygen boiling point of −182.97°C, and the junction of the platinum resistance thermometer range and the thermocouple range became the measured antimony freezing point (about 630°C) in place of 660°C; the silver freezing point was defined as being 960.8°C instead of 960.5°C; the gold freezing point replaced the gold melting point (1063°C); the Planck radiation law replaced the Wien law; the value assigned to the second radiation constant became 1.438×10^{-2} m·K in place of 1.432×10^{-2} m·K; the permitted ranges for the constants of the interpolation formulae for the standard resistance thermometer and thermocouple were modified; the limitation of T for optical pyrometry ($\lambda T < 3 \times 10^{-3}$ m·K) was changed to the requirement that 'visible' radiation be used.

9.4.3 International Practical Temperature Scale of 1948 (IPTS–48) (amended edition of 1960)

The International Practical Temperature Scale of 1948, amended edition of 1960, was adopted by the eleventh General Conference: the tenth General Conference had already adopted the triple point of water as the sole point defining the kelvin, the unit of thermodynamic temperature. In addition to the introduction of the word 'Practical', the modifications to the ITS–48 were: the triple point of water, defined as being 0.01 °C, replaced the melting point of ice as the calibration point in this region; the freezing point of zinc, defined as being 419.505 °C, became a preferred alternative to the sulphur boiling point (444.6 °C) as a calibration point; the permitted ranges for the constants of the interpolation formulae for the standard resistance thermometer and the thermocouple were further modified; the restriction to 'visible' radiation for optical pyrometry was removed.

Inasmuch as the numerical values of temperature on the IPTS–48 were the same as on the ITS–48, the former was not a revision of the scale of 1948 but merely an amended form of it.

9.4.4 International Practical Temperature Scale of 1968 (IPTS–68)

In 1968 the International Committee of Weights and Measures promulgated the International Practical Temperature Scale of 1968, having been empowered to do so by the thirteenth General Conference of 1967–1968. The IPTS–68 incorporated very extensive changes from the IPTS–48. These included numerical changes, designed to bring it more nearly in accord with thermodynamic temperatures, that were sufficiently large to be apparent to many users. Other changes were as follows: the lower limit of the scale was extended down to 13.81 K; at even lower temperatures (0.5 K to 5.2 K), the use of two helium vapour pressure scales was recommended; six new defining fixed points were introduced — the triple point of equilibrium hydrogen (13.81 K), an intermediate equilibrium hydrogen point (17.042 K), the normal boiling point of equilibrium hydrogen (20.28 K), the boiling point of neon (27.102 K), the triple point of oxygen (54.361 K), and the freezing point of tin (231.9681 °C) which became a permitted alternative to the boiling point of water; the boiling point of sulphur was deleted; the values assigned to four fixed points were changed — the boiling point of oxygen (90.188 K), the freezing point of zinc (419.58 °C), the freezing point of silver (961.93 °C), and the freezing point of gold (1064.43 °C); the interpolating formulae for the resistance thermometer range became much more complex; the value assigned to the second radiation constant c_2 became 1.4388×10^{-2} m·K; the permitted ranges of the constants for the interpolation formulae for the resistance thermometer and thermocouple were again modified.

9.4.5 International Practical Temperature Scale of 1968 (IPTS–68) (amended edition of 1975)

The International Practical Temperature Scale of 1968, amended edition of 1975, was adopted by the fifteenth General Conference in 1975. As was the case for the IPTS–48 with respect to the ITS–48, the IPTS–68(75) introduced no numerical changes. Most of the extensive textual changes were intended only to clarify and simplify its use. More substantive changes were: the oxygen point was defined as the condensation point rather than the boiling point; the triple point of argon (83.798 K) was introduced as a permitted alternative to the condensation point of oxygen; new values of the isotopic composition of naturally occurring neon were adopted; the recommendation to use values of T given by the 1958 ^{4}He and 1962 ^{3}He vapour-pressure scales was rescinded.

9.4.6 1976 Provisional 0.5 to 30 K Temperature Scale (EPT–76)

The 1976 Provisional 0.5 K to 30 K Temperature Scale was introduced to meet two important requirements: these were to provide means of substantially reducing the errors (with respect to corresponding thermodynamic values) below 27 K that were then known to exist in the IPTS–68 and throughout the temperature ranges of the ^{4}He and ^{3}He vapour pressure scales of 1958 and 1962 respectively, and to bridge the gap between 5.2 K and 13.81 K in which there had not previously been an international scale. Other objectives in devising the EPT–76 were 'that it should be thermodynamically smooth, that it should be continuous with the IPTS–68 at 27.1 K, and that it should agree with thermodynamic temperature T as closely as these two conditions allow'. In contrast with the IPTS–68, and to ensure its rapid adoption, several methods of realising the EPT–76 were approved. These included: using a thermodynamic interpolation instrument and one or more of eleven assigned reference points; taking differences from the IPTS–68 above 13.81 K; taking differences from helium vapour pressure scales below 5 K; and taking differences from certain well-established laboratory scales. Because there was a certain 'lack of internal consistency' it was admitted that 'slight ambiguities between realisations' might be introduced. However the advantages gained by adopting the EPT–76 as a working scale until such time as the IPTS–68 should be revised and extended were considered to outweigh the disadvantages.

9.4.7   International Temperature Scale of 1990 (ITS–90)

This present text, ITS–90, was adopted by the International Committee of Weights and Measures at its meeting in 1989. It incorporated very extensive changes from IPTS–68. These included numerical changes designed to bring it more nearly in accord with thermodynamic temperatures. The ITS–90 comprises a number of ranges and sub-ranges throughout which temperatures T_{90} are defined. Several of the ranges or sub-ranges overlap and where over-lapping occurs the different definitions of T_{90} have equal status. The lower limit of the scale was extended to 0.65 K and helium vapour pressure relationships and interpolating gas thermometry used in the sub-ranges up to 24.5561 K. The platinum resistance thermometer was defined as the interpolating instrument from 13.8033 K to 961.78 °C and two reference functions were defined from 13.8033 to 273.16 K, and from 0 °C to 961.78 °C. The Pt 10% Rh-vs-Pt thermocouple was removed as an interpolating instrument and realisation of the scale using the Planck radiation law was lowered to 961.78 °C.

9.5 Supplementary information and differences from earlier scales

The apparatus, methods and procedures that will serve to realise the ITS–90 are given in reference [20]. This document also gives an account of the earlier International Temperature Scales and the numerical differences between successive scales, which include, where practicable, mathematical functions for the differences $T_{90} - T_{68}$. Also, reference [49] contains a number of useful approximations to the ITS–90. Both documents [20, 49] were prepared by the Comité Consultatif de Thermométrie and are published by the BIPM; they are revised and updated periodically.

The difference $T_{90} - T_{68}$ is given as a function of temperature in Table 9.6— the values are from reference [48] with the data from 631 to 1064 °C amended using the polynomial of reference [50]. The number of significant digits used in the table allows smooth interpolations to be made. However, the reproducibility of the IPTS–68 is, in many areas, substantially worse than is implied by this number.

Table 9.6: Differences between ITS–90 and EPT–76, in mK, and between ITS–90 and IPTS–68, in K (or °C).

$(T_{90} - T_{76})$:

T_{90} (K)	0	1	2	3	4	5	6	7	8	9
0						-0.1	-0.2	-0.3	-0.4	-0.5
10	-0.6	-0.7	-0.8	-1.0	-1.1	-1.3	-1.4	-1.6	-1.8	-2.0
20	-2.2	-2.5	-2.7	-3.0	-3.2	-3.5	-3.8	-4.1		

$(T_{90} - T_{68})$:

T_{90} (K)	0	1	2	3	4	5	6	7	8	9
10					-0.006	-0.003	-0.004	-0.006	-0.008	-0.009
20	-0.009	-0.008	-0.007	-0.007	-0.006	-0.005	-0.004	-0.004	-0.005	-0.006
30	-0.006	-0.007	-0.008	-0.008	-0.008	-0.007	-0.007	-0.007	-0.006	-0.006
40	-0.006	-0.006	-0.006	-0.006	-0.006	-0.007	-0.007	-0.007	-0.006	-0.006
50	-0.006	-0.005	-0.005	-0.004	-0.003	-0.002	-0.001	0.000	0.001	0.002
60	0.003	0.003	0.004	0.004	0.005	0.005	0.006	0.006	0.007	0.007
70	0.007	0.007	0.007	0.007	0.007	0.008	0.008	0.008	0.008	0.008
80	0.008	0.008	0.008	0.008	0.008	0.008	0.008	0.008	0.008	0.008
90	0.008	0.008	0.008	0.008	0.008	0.008	0.008	0.009	0.009	0.009

T_{90} (K)	0	10	20	30	40	50	60	70	80	90
100	0.009	0.011	0.013	0.014	0.014	0.014	0.014	0.013	0.012	0.012
200	0.011	0.010	0.009	0.008	0.007	0.005	0.003	0.001		

$(t_{90} - t_{68})$:

t_{90} (°C)	0	-10	-20	-30	-40	-50	-60	-70	-80	-90
-100	0.013	0.013	0.014	0.014	0.014	0.013	0.012	0.010	0.008	0.008
0	0.000	0.002	0.004	0.006	0.008	0.009	0.010	0.011	0.012	0.012

t_{90} (°C)	0	10	20	30	40	50	60	70	80	90
0	0.000	-0.002	-0.005	-0.007	-0.010	-0.013	-0.016	-0.018	-0.021	-0.024
100	-0.026	-0.028	-0.030	-0.032	-0.034	-0.036	-0.037	-0.038	-0.039	-0.039
200	-0.040	-0.040	-0.040	-0.040	-0.040	-0.040	-0.040	-0.039	-0.039	-0.039
300	-0.039	-0.039	-0.039	-0.040	-0.040	-0.041	-0.042	-0.043	-0.045	-0.046
400	-0.048	-0.051	-0.053	-0.056	-0.059	-0.062	-0.065	-0.068	-0.072	-0.075
500	-0.079	-0.083	-0.087	-0.090	-0.094	-0.098	-0.101	-0.105	-0.108	-0.112
600	-0.115	-0.118	-0.122	-0.125	-0.11	-0.10	-0.09	-0.07	-0.05	-0.04
700	-0.02	-0.01	0.00	0.02	0.03	0.03	0.04	0.05	0.05	0.05
800	0.05	0.05	0.04	0.04	0.03	0.02	0.01	0.00	-0.02	-0.03
900	-0.05	-0.06	-0.08	-0.10	-0.11	-0.13	-0.15	-0.16	-0.18	-0.19
1000	-0.20	-0.22	-0.23	-0.23	-0.24	-0.25	-0.25	-0.25	-0.26	-0.26

t_{90} (°C)	0	100	200	300	400	500	600	700	800	900
1000		-0.26	-0.30	-0.35	-0.39	-0.44	-0.49	-0.54	-0.60	-0.66
2000	-0.72	-0.79	-0.85	-0.93	-1.00	-1.07	-1.15	-1.24	-1.32	-1.41
3000	-1.50	-1.59	-1.69	-1.78	-1.89	-1.99	-2.10	-2.21	-2.32	-2.43

Bibliography

[1] G. Rosengarten. Physics of temperature measurement. In R. E. Bentley, editor, *Temperature and Humidity Measurements*, volume 1 of *Handbook of Temperature Measurement*, chapter 2. Springer-Verlag, Singapore, 1998.

[2] R. C. Kemp, L. M. Besley, and W. R. G. Kemp. Constant volume gas thermometry from 13.8 to 83.8 K. In J. F. Schooley, editor, *Temperature: its Measurement and Control in Science and Industry*, volume 5, pages 33–38. Am. Inst. Physics, New York, 1982.

[3] L. A. Guildner and R. E. Edsinger. Deviation of international practical temperatures from thermodynamic temperatures in the temperature range from 273.16 to 730 k. *J. Res. Natl Bur. Stds–Sect. A*, 80A:703–38, 1976.

[4] R. E. Edsinger and J. F. Schooley. Differences between thermodynamic temperatures and t(IPTS–68) in the range 230 to 660°C. *Metrologia*, 26:95–106, 1989.

[5] T. J. Quinn and J. E. Martin. A black-body cavity for total-radiation thermometry. *Metrologia*, 23:111–14, 1986.

[6] H. J. Jung. A measurement of thermodynamic temperatures between 683 K and 933 K by infrared pyrometry. *Metrologia*, 23:19–31, 1986.

[7] J. J. Connolly, T. P. Jones, and J. Tapping. Comparison of resistance versus thermodynamic temperature of platinum resistance thermometers with ITS-90. *Metrologia*, 27:83–88, 1990.

[8] L. M. Besley. Cryogenic thermometry. In R. E. Bentley, editor, *Temperature and Humidity Measurements*, volume 1 of *Handbook of Temperature Measurement*, chapter 3. Springer-Verlag, Singapore, 1998.

[9] H. L. Callendar and E. H. Griffiths. On the determination of the boiling point of sulphur and on a method of standardizing platinum resistance

thermometers by reference to it—experiments made at the Cavendish Laboratory. *Phil. Trans. Roy. Soc.*, A182:119–57, 1891.

[10] H. L. Callendar. On a practical thermometric standard. *Phil. Mag.*, 48:519–47, 1899.

[11] C. H. Meyers. Coiled filament resistance thermometers. *Natl Bur. Stds J. Res.*, 9:807, 1932.

[12] N. M. Bass, J. P. Evans, and S. B. Tillett. Techniques in high-temperature platinum resistance thermometry. NBS Tech. Note 1183, Natl Bur. Stds, USA, 1984.

[13] L. Guang and T. Hongtu. Stability of precision high-temperature platinum resistance thermometers. In J. F. Schooley, editor, *Temperature: its Measurement and Control in Science and Industry*, volume 5, pages 783–87. Am. Inst. Physics, New York, 1982.

[14] P. Marcarino, R. Dematteis, M. Gallorini, and E. Rizzio. Contamination of PRT's at high temperatures through their silica sheaths. *Metrologia*, 26:175–181, 1989.

[15] J. P. Evans and D. M. Sweger. Immersion cooler for freezing ice mantles on triple-point-of-water cells. *Rev. Sci. Instrum.*, 40:376–77, 1969.

[16] E. H. McLaren. The freezing points of high-purity metals as precision temperature standards. In F. G. Brickwedde, editor, *Temperature: its Measurement and Control in Science and Industry, vol. 3*, pages 185–98. Rheinhold, New York, 1962.

[17] J. V. McAllan and M. M. Ammar. Comparison of the freezing points of aluminium and antimony. In *Temperature: its Measurement and Control in Science and Industry*, volume 4, pages 275–85. Inst. Soc. America, Pittsburgh, 1972.

[18] B. W. Mangum. Determination of the indium freezing-point and triple-point temperatures. *Metrologia*, 26:211–17, 1989.

[19] B. W. Mangum and D. D. Thornton. Determination of the triple-point temperature of gallium. *Metrologia*, 15:201–15, 1979.

[20] Comité Consultatif de Thermométrie. Supplementary information for the International Temperature Scale of 1990. BIPM monograph, Pavillon de Breteuil, F-92310 Sèvres, France, 1990.

[21] *ISO Guide to the Expression of Uncertainty in Measurement*. International Organization for Standardization, Geneva, 1993.

[22] R. E. Bentley. The uncertainty in temperature measurement. In *Theory and Practice of Thermoelectric Thermometry*, volume 3 of *Handbook of Temperature Measurement*, chapter 5. Springer-Verlag, Singapore, 1998.

[23] IEC-751. Industrial platinum resistance thermometer sensors. Technical report, International Electrochemical Commission, Genève, Suisse, 1983.

[24] N. M. Bass and J. J. Connolly. The performance of industrial platinum resistance thermometers. *Aust. J. Instrum. & Control*, 36:88–90, 1980.

[25] M. V. Chattle. Resistance ratio/temperature relationships for industrial resistance thermometers. Technical report, NPL report qu30, National Physical Laboratory, U.K., 1975.

[26] R. E. Bentley. The calibration of thermocouples. In *Theory and Practice of Thermoelectric Thermometry*, volume 3 of *Handbook of Temperature Measurement*, chapter 4. Springer-Verlag, Singapore, 1998.

[27] J. J. Connolly. The gas-pressure-controlled caesium heat-pipe as a thermometer-calibration enclosure. In *Proc. Int. Conf. on Meas. Science, Tech. and Practice*, Melbourne, Australia, November 1997.

[28] E. F. Mueller. Bridges for resistance thermometry. *Bull. Natl Bur. Stds*, 13:547–561, 1917.

[29] F. E. Smith. On bridge methods for resistance measurement of high precision in platinum thermometry. *Phil. Mag.*, 24:541–569, 1912.

[30] J. J. Connolly, J. V. McAllan, and G. W. Small. Resistance thermometry using a new design of AC bridge. In *Temperature: its Measurement and Control in Science and Industry, vol. 4*, pages 1487–93, 1972.

[31] D. R. White and J. M. Williams. Resistance network for verifying the accuracy of resistance bridges. *IEEE Trans. Instrum. Meas.*, 46:329–32, 1997.

[32] R. E. Bentley. *Theory and Practice of Thermoelectric Thermometry*, volume 3 of *Handbook of Temperature Measurement*. Springer-Verlag, Singapore, 1998.

[33] W. E. K. Middleton. *A History of the Thermometer and its Use in Meteorology*. Johns Hopkins Press, Baltimore, 1968.

[34] AS 2831. Thermometers–solid stem–long and short–for precision use. Australian Standard, Standards Australia, Sydney, 1985.

[35] J. A. Wise and R. J. Soulen. Thermometer calibration: a model for state calibration laboratories. Monogr. 174, Natl. Bur. Stds, 1986.

[36] AS 2819. Thermometers—meteorological—maximum, minimum and ordinary. Australian Standard, Standards Australia, Sydney, 1985.

[37] AS 2190. Clinical maximum thermometers. Australian Standard, Standards Australia, Sydney, 1978.

[38] G. Sandars and M. J. Ballico. Traceable measurements. In R. E. Bentley, editor, *Temperature and Humidity Measurements*, volume 1 of *Handbook of Temperature Measurement*, chapter 1. Springer-Verlag, Singapore, 1998.

[39] T. J. Quinn. *Temperature*. Monographs in Physical Measurement. Academic Press, London, 1983.

[40] J. L. Riddle, G. T. Furukawa, and H. H. Plumb. Platinum resistance thermometry. Monogr. 126, Natl. Bur. Stds, 1972.

[41] J. Tapping. Temperature control. In R. E. Bentley, editor, *Temperature and Humidity Measurements*, volume 1 of *Handbook of Temperature Measurement*, chapter 10. Springer-Verlag, Singapore, 1998.

[42] T. P. Jones, S. R. Meszaros, and T. M. Morgan. Design and construction of thermocouple calibration furnaces. *Aust. Journal of Instrumentation and Control*, pages 106–11, 1982.

[43] R. E. Bentley and S. R. Meszaros. A laboratory diagnostic furnace for scanning and calibrating thermocouples. *Aust. Journal of Instrumentation and Control*, 4(4):4–8, 1989.

[44] C. Bassani, C. A. Busse, and F. Geiger. High-precision heat-pipe furnaces. *High Temp.-High Pressures*, 12:351–56, 1979.

[45] G. K. White. *Experimental Techniques in Low-Temperature Physics*. Oxford University Press, Oxford, 3rd edition, 1979.

[46] R. E. Bentley. Multi-site temperature measurement. In *Theory and Practice of Thermoelectric Thermometry*, volume 3 of *Handbook of Temperature Measurement*, chapter 6. Springer-Verlag, Singapore, 1998.

[47] R. E. Bentley. Thermocouples in use. In *Theory and Practice of Thermoelectric Thermometry*, volume 3 of *Handbook of Temperature Measurement*, chapter 3. Springer-Verlag, Singapore, 1998.

[48] H. Preston-Thomas. The International Temperature Scale of 1990 (ITS–90). *Metrologia*, 27:3–10, 1990.

[49] Comité Consultatif de Thermométrie. Techniques for approximating the International Temperature Scale of 1990. BIPM monograph, Pavillon de Breteuil, F-92310 Sèvres, France, 1990.

[50] G. W. Burns, G. F. Strouse, B. W. Mangum, M. C. Croarkin, W. F. Guthrie, P. Marcarino, M. Battuello, H. K. Lee, J. C. Kim, K. S. Gam, C. Rhee, M. Chattle, M. Arai, H. Sakurai, A. I. Pokhodun, N. P. Moiseeva, S. A. Perevalova, M. J. de Groot, J. Zhang, K. Fan, and S. Wu. New reference functions for platinum-10% rhodium versus platinum (type S) thermocouples based on the ITS–90. In J. F. Schooley, editor, *Temperature: its Measurement and Control in Science and Industry, Vol. 6*, pages 537–46. Amer. Inst. Phys., New York, 1992.